Military Doctrine

Recent Titles in
Contemporary Military, Strategic, and Security Issues

Military Reform: A Reference Handbook
Winslow T. Wheeler and Lawrence J. Korb

The U.S. Military's "Don't Ask, Don't Tell" Policy: A Reference Handbook
Melissa Sheridan Embser-Herbert

Prisoners of War: A Reference Handbook
Arnold Krammer

Nation-Building and Stability Operations: A Reference Handbook
Cynthia A. Watson

Military Transformation and Modern Warfare: A Reference Handbook
Elinor Sloan

Information Operations—Doctrine and Practice: A Reference Handbook
Christopher Paul

The National Guard and Reserve: A Reference Handbook
Michael D. Doubler

Returning Wars' Wounded, Injured, and Ill: A Reference Handbook
Nathan D. Ainspan and Walter E. Penk, editors

Manning the Future Legions of the United States: Finding and Developing Tomorrow's Centurions
Donald Vandergriff

The Process and Politics of Defense Acquisition: A Reference Handbook
David S. Sorenson

International Crime and Punishment: A Guide to the Issues
James Larry Taulbee

Serving America's Veterans: A Reference Handbook
Lawrence J. Korb, Sean E. Duggan, Peter M. Juul, and Max A. Bergmann

Military Doctrine

A Reference Handbook

Bert Chapman

Contemporary Military, Strategic, and Security Issues

Praeger Security International
An Imprint of ABC-CLIO, LLC

A B C CLIO

Santa Barbara, California • Denver, Colorado • Oxford, England

Library of Congress Cataloging-in-Publication Data

Chapman, Bert.
Military doctrine : a reference handbook / Bert Chapman.
p. cm.
Includes bibliographical references and indexes.
ISBN 978-0-313-35233-1 (hardcover : acid-free paper) — ISBN 978-0-313-35234-8 (ebook)
1. Military doctrine—United States—Handbooks, manuals, etc. 2. Military doctrine—Handbooks, manuals, etc. 3. Deployment (Strategy)—Handbooks, manuals, etc. 4. Combat—Handbooks, manuals, etc. 5. Logistics—Handbooks, manuals, etc. 6. Tactics—Handbooks, manuals, etc. 7. United States—Military policy. I. Title.
UA23.C5134 2009
355'.033073—dc22 2009016484

13 12 11 10 9 1 2 3 4 5

This book is also available on the World Wide Web as an eBook.
Visit www.abc-clio.com for details.

ABC-CLIO, LLC
130 Cremona Drive, P.O. Box 1911
Santa Barbara, California 93116-1911

This book is printed on acid-free paper ∞

Manufactured in the United States of America

To

my parents,

Albert and Mildred Chapman,

and

my brother,

Brent Chapman, for their love,

direction, encouragement, and support.

Contents

Acknowledgments

Numerous individuals have contributed to this book's appearance. At Purdue University Libraries, I am blessed to work with librarians who encourage scholarly excellence. Particularly helpful guidance has been provided by my Purdue colleague, Jean-Pierre Herubel, whose encyclopedic mastery of scholarly publishing practices has contributed to the chapter on grey literature. Purdue's Interlibrary Loan Department has also provided access to source material not available in locally owned or accessible print or digital collections. This work has been made much easier by the widespread Internet availability of military doctrine and national security strategy documents from the United States and a number of foreign governments and militaries.

The high quality work of Lori Bryant and Libby Wahl of the Government Documents Department support staff, as well as of student workers, such as Megan Cochran, made it possible for me to have the time to write such a work.

Steve Catalano, Tim Furnish, Adam Kane, and Heather Ruland Staines have all helped guide me through PSI's publishing practices and procedures with consummate professionalism.

I am especially blessed by the love and support provided by my wife Becky throughout all of life's circumstances. I also want to acknowledge the support and encouragement that my parents and brother have always given me.

Introduction

Napoleon Bonaparte's declaration that an army marches on its stomach is a classic military history axiom.[1] However, while this saying may be true in a nutritive or logistical sense, it takes far more than physical nourishment to enable military forces to conduct and sustain their operations. More substantive, intellectual fiber is required for these forces to launch, sustain, and conclude their operations, and this cerebral foundation is called military doctrine.

Military doctrine can and has been defined in many different ways in numerous countries. These varying definitions are affected by the security factors that face these countries, as well as existing and emerging technological trends and developments, internal political factors within the armed services, such as interservice competition and conflicting perspectives of civilian and military policymakers regarding critical national security priorities, and budgetary factors that may compel armed services to downsize their military objectives.

One appraisal of military doctrine describes it as focusing military strategic capabilities to determine strategic objectives and desired final results, detailing required military action, allocating resources, and restraining such allocations as directed by political leaders.[2] Another assessment from the early 1990s asserts that military doctrine has three different emphases:

- Guaranteeing security at the expense of other countries and reducing overall security;
- Guaranteeing national security by equalizing a threat and stabilizing overall security;
- Guaranteeing national security by increasing other countries' sense of security, consequently weakening sources of threat.[3]

A recent British assessment provides the following definition of military doctrine, which encompasses its interdisciplinary breadth with considerable succinctness and bracing clarity:

> Military forces have among other things the distinctive ability to use combat. They are in the business of the organized use of violence. The study of combat

> embraces a large number of intellectual Disciplines spanning the exact sciences such as physics on the one extreme to the liberal arts such as history on the other. Combat itself creates and exploits havoc and, as Clausewitz warned us, the onset of combat makes for uncertainty of outcome however good the planning. Doctrine provides the intellectual structure for the practitioners, military commanders at every level and their staffs and subordinates, to think sensibly about the application of military force and to be guided by sound reasoning.[4]

This appraisal goes on to contend that the writing of military doctrine is a simplifying process—a product of intellectual activity to determine how military force should be applied. It stresses that individual armed service branches will disagree over how prescriptive doctrine should be, and some service doctrine writers will believe that military doctrine should present solutions instead of options. A summative aspect of such doctrine is that it provides a coherent and consistent framework of concepts, tenets, and principles that are applicable in planning and conducting operations, and that these doctrinal attributes are intended to assist in developing and executing operational plans.[5]

Militaries have sought to develop rational, scientific means for formulating, documenting, and justifying their military policies to pursue objectives they define as being in their national security interests. A significant body of literature documents the justifications for these national military doctrines. This literature encompasses countries like Germany, which can arguably be considered the originator of national military doctrine.[6]

Great Britain was Germany's rival in formulating a coherent body of military doctrine that encompassed multiple armed service branches.[7] The former Soviet Union and the Russian Federation have also made significant contributions to military doctrine, and a considerable body of literature that analyzes Russian military viewpoints is available.[8]

The United States has been the biggest producer of military doctrine documentation, and it has received significant and substantive scrutiny.[9] China's growing economic wealth has also prompted it to invest additional resources in its military. There exist varied assessments of whether China's military is or will become a threat to the United States, and there is also a steadily growing body of scholarly literature examining historical and contemporary Chinese military doctrine and what it may mean for future Chinese military action.[10]

Other countries, including Australia,[11] Canada,[12] India,[13] Israel,[14] and South Africa,[15] have also crafted and developed military doctrine to inform their policies for conducting military operations. This literature will also receive scrutiny in this book.

Military Doctrine: A Reference Handbook serves as an introductory overview to the role military doctrine has played and will continue to play in the development of national military policy, and it provides a detailed overview of documentary and scholarly literature from the United States and other countries.

This book begins by describing key events in the post–World War II military doctrinal history of the United States and other countries, and then considers possible developments in the military doctrine of these countries.

Subsequent chapters describe military doctrinal publications produced by the United States and other countries, as well as how to find these publications on the Internet. These chapters also examine such publications to learn more about the military doctrinal policies of these countries. Additionally, this book reviews scholarly literature and grey literature, such as dissertations, that describe and analyze military doctrine.

It is important to study military doctrine in order to understand how and why countries have conducted military operations in the past, as well as why they currently engage in such operations and how they may conduct them in the future. Documents describing military doctrine cover the various aspects of land, air, and naval warfare, intelligence operations, peacekeeping operations, information warfare, and the nascent military arena of space. This literature will not always reflect how military forces actually conduct combat operations or how evolving battlefield, domestic political, and international diplomatic realities may compel changes in military doctrine and operational conduct. This literature will, however, reflect the basic intellectual, cultural, normative, and political foundations motivating national decisions to conduct operations against other countries or terrorist organizations.

Consequently, this work will be most beneficial to military officers, historians, political scientists, and students of military history and national security policymaking who desire to enhance their understanding of the historical, contemporary, and future importance of military doctrinal literature in domestic and international military policymaking.

Notes

1. Richard Glover, "War and Civilian Historians," *Journal of the History of Ideas* 18 (1957): 91.

2. G. L. Garnett, "The Evolution of the Canadian Approach to Joint and Combined Operations at the Strategic and Operational Level," *Canadian Military Journal* 3, no. 4 (2002–2003): 6.

3. Stanislaw Koziej, "Pan-European Security System: Future Military Doctrine?," *Military Review* 72, no. 12 (1992): 48–49.

4. Michael Codner, "Purple Prose and Purple Passion: The Joint Defence Centre," *RUSI Journal* 144 (1999): 37.

5. Ibid.

6. For a partial monographic sampling of this literature, see Gordon A. Craig, *The Politics of the Prussian Army, 1640–1945* (Oxford: Clarendon Press, 1955); Jehuda Lothar Wallach, *The Dogma of the Battle of Annihilation: The Theories of Clausewitz and Schlieffen and Their Impact on the German Conduct of Two World Wars* (Westport, CT: Greenwood Press, 1986); James S. Corum, *Roots of Blitzkrieg: Hans von Seeckt and German Military Reform* (Lawrence: University Press of Kansas, 1992); Antulio J. Echevarria II, *After Clausewitz: German*

Military Thinkers before the Great War (Lawrence: University Press of Kansas, 2000); Robert M. Citino, *The Path to Blitzkrieg: Doctrine and Training in the German Army, 1920–1939* (Boulder, CO: Lynne Rienner Publishers, 1999); Mary R. Habeck, *Storm of Steel: The Development of Armor Doctrine in Germany and the Soviet Union, 1919–1939* (Ithaca, NY: Cornell University Press, 2003); and Robert M. Citino, *The German Way of War: From the Thirty Years War to the Third Reich* (Lawrence: University Press of Kansas, 2005).

7. Barry Posen, *The Sources of Military Doctrine: France, Britain, and Germany between the World Wars* (Ithaca, NY: Cornell University Press, 1984); Martin Samuels, *Command or Control?: Command, Training, and Tactics in the British and German Armies, 1888–1918* (London: Frank Cass, 1995); Elizabeth Kier, *Imagining War: French and British Military Doctrine between the Wars* (Princeton, NJ: Princeton University Press, 1997); John Stone, *The Tank Debate: Armour and the Anglo-American Military Tradition* (Amsterdam: Harwood Academic Publishers, 2000); and M. A. Ramsay, *Command and Cohesion: The Citizen Soldier and Minor Tactics in the British Army, 1870–1918* (Westport, CT: Praeger, 2002).

8. See Habeck, *Storm of Steel;* Bruce W. Menning, *Bayonets before Bullets: The Imperial Russian Army, 1861–1914* (Bloomington: Indiana University Press, 1992); Willard C. Frank, Jr. and Philip S. Gillette, eds., *Soviet Military Doctrine from Lenin to Gorbachev, 1915–1991* (Westport, CT: Greenwood Press, 1992). Jonathan Samuel Lockwood, *Russian View of U.S. Strategy: Its Past, Its Future* (New Brunswick, NJ: Transaction, 1993); James H. Slagle, "New Russian Military Doctrine: Sign of the Times," *Parameters* 24 (1994): 88–99; Stephen J. Blank, *Russian Armed Forces on the Brink of Reform* (Carlisle Barracks, PA: Strategic Studies Institute, U.S. Army War College, 1998); and James Sterrett, *Soviet Air Force Theory, 1918–1945* (London: Routledge, 2007).

9. See, for example, Colin S. Gray, *Weapons Don't Make War: Policy, Strategy, and Military Technology* (Lawrence: University Press of Kansas, 1993); John L. Romjue, *American Army Doctrine for the Post–Cold War* (Fort Monroe, VA: Military History Office, United States Army Training and Doctrine Command, 1996); Andrew James Birtle, *U.S. Army Counterinsurgency and Contingency Operations Doctrine, 1860–1941* (Washington, DC: U.S. Army Center of Military History, 1998); Kenneth Finlayson, *Uncertain Trumpet: The Evolution of U.S. Army Infantry Doctrine, 1919–1941* (Westport, CT: Greenwood Press, 2001); Michael A. Vane and Robert M. Toguchi, *The Enduring Relevance of Landpower: Flexibility and Adaptability for Joint Campaigns* (Arlington, VA: Institute of Land Warfare, Association of the United States Army, 2003); and Rudolph M. Janiczek, *A Concept at the Crossroads: Rethinking the Center of Gravity* (Carlisle Barracks, PA: Strategic Studies Institute, U.S. Army War College, 2007), http://purl.access.gpo.gov/GPO/LPS87489.

10. See Karl W. Eikenberry, "Does China Threaten Asia-Pacific Regional Stability?" *Parameters* 25 (1995): 82–103; Alistair Iain Johnston, *Cultural Realism: Strategic Culture and Grand Strategy in Chinese History* (Princeton, NJ: Princeton University Press, 1995); John Hill, "China's Military Modernization Takes Shape," *Jane's Intelligence Review* 16, no. 2 (2004): 46–50; Ka-po Ng, *Interpreting China's Military Power: Doctrine Makes Readiness* (London: Frank Cass, 2005); Roger Cliff, Michael S. Chase, Derek Eaton, and Kevin L. Pollpeter, *Entering the Dragon's Lair: Chinese Anti-Access Strategies and Their Implications for the United States* (Santa Monica, CA: Rand Project Air Force, 2007).

11. Representative examples of a rich literary corpus on Australian military doctrine include Mark Christopher John Welburn, *The Development of Australian Army Doctrine, 1945–1964* (Canberra: Strategic and Defence Studies Centre, Research School of Pacific and Asian Studies, Australian National University, 1994); C. J. Shine, "Restructuring the Australian Army: The Seeds of Future Crisis?," *Australian Defence Force Journal* 131 (1998): 5–17;

Alan Ryan, "The Challenge of 'New Times': Developing Doctrine for an Uncertain Future," *Australian Defence Force Journal* 142 (2000): 49–54; Michael Evans, ed., *Changing the Army: The Roles of Doctrine, Development and Training* (Canberra: Land Warfare Studies Centre, 2000); and Michael Evans, *The Tyranny of Dissonance: Australia's Strategic Culture and Way of War, 1901–2005* (Duntroon: Land Warfare Studies Centre, 2005).

12. See J. W. Hammond, *First Things First: Improving Canadian Leadership Doctrine* (Toronto: Canadian Forces Command Staff College, 1996); R. K. Taylor, "2020 Vision: Canadian Forces Operational-Level Doctrine," *Canadian Military Journal* 2, no. 3 (2001): 35–42; G. L. Garnett, "Evolution of the Canadian Approach," 6; and Paul Grimshaw, *Conduct after Capture and Terrorist Hostage Taking: A Case for New Doctrine* (Toronto: Canadian Forces College, 2007).

13. P. K. Chakravorty, "Artillery Revolution: An Indian Perspective," *Military Technology* 28, no. 7 (2004): 81–83; Lowell Dittmer, ed., *South Asia's Nuclear Security Dilemma: India, Pakistan, and China* (Armonk, NY: M.E. Sharpe, 2005); and Harsh V. Pant, "India's Nuclear Doctrine and Command Structure: Implications for Civil-Military Relations in India," *Armed Forces & Society* 33 (2007): 238–264.

14. Frank K. Sobchak, "'Ah Harey'—Follow Me—Origins of the Israeli Junior Leadership Doctrine," *Military Intelligence* 19, no. 4 (1993): 20–23; Gabriel Ben-Dor, Ami Pedahzur, and Badi Hasisi, "Israel's National Security Doctrine under Strain: The Crisis of the Reserve Army," *Armed Forces & Society* 28 (2002): 233–255; Sergio Catagnani, "Israel Defence Forces Organizational Changes in an Era of Budgetary Cutbacks," *RUSI Journal* 149, no. 5 (2004): 72–76; and Uri Bar-Joseph, "The Paradox of Israeli Power," *Survival* 46, no. 4 (2004–2005): 137–156.

15. Dean Fourie, "South Africa's Developing Security and Defence Policies," *RUSI Journal* 135, no. 2 (1990): 25–30; Chris Bennett, "No Room for 'Nice to Haves'," *U.S. Naval Institute Proceedings* 126, no. 3 (2000): 44–47; M. Hough and L. Du Pessis, eds., *Selected Military Issues with Specific Reference to the Republic of South Africa* (Pretoria: Institute for Strategic Studies, University of Pretoria, 2001); and "South Africa's New Defence Strategy," *Military Technology* 30 (2006): 284–286.

CHAPTER 1

U.S. Military Doctrine: A Selective Post–World War II History

The six decades since World War II have seen tremendous developments and changes in U.S. military doctrine. These changes have influenced and continue to influence this doctrine as U.S. military leaders and civilian national security policymakers have sought to develop and implement military strategy and doctrine to enable U.S. military forces to achieve desired national objectives.

U.S. military doctrine encompasses conventional military operations, potential military operations involving nuclear weapons and other weapons of mass destruction, and unconventional means of warfare, such as counterinsurgency, peacekeeping, and humanitarian operations.

This chapter seeks to provide a selective overview of major U.S. military doctrine developments from World War II to the present. It does not aspire to be a comprehensive history of U.S. military doctrine during this time period. U.S. military historians and scholars of U.S. military doctrinal development may not agree with the importance of the doctrinal developments highlighted in this chapter. It is hoped that this chapter will give those readers interested in U.S. military doctrine a representative sampling and substantive introductory overview to some of the most critical events in post–World War II U.S. military doctrinal trends and development. Such an overview will, hopefully, pique readers' desire to learn more about U.S. doctrinal development, as well as the military doctrinal development of other countries and international governmental organizations, and provide a comprehensive understanding of how to conduct substantive scholarly research on military doctrine using the field's primary and secondary sources of literature.

Soon after the successful conclusion of World War II, the United States' relations with its wartime ally the Soviet Union began to deteriorate for various political, ideological, and military strategic reasons, and a Cold War developed. This conflict would last for over four and a half decades and profoundly influence U.S. military doctrine, national security strategy, and foreign policy. Recognition of the long-term nature of the United States' rivalry with the Soviet Union resulted in the development of *NSC-68: United States Objectives and Programs for National Security*—a key U.S. national security strategy document created in 1950.

Written by individuals such as Paul Nitze and often viewed as a strategic companion to George Kennan's "Long Telegram," which had a similar focus, *NSC-68* maintained that the United States needed to increase its military strength to counter a fanatical Soviet ideology that sought to impose itself on the world. This document went on to add that the United States should build an international community and pursue a containment strategy that would seek to prevent further Soviet Communist advances by emphasizing military instead of diplomatic action and pursuing policies of calculated and gradual coercion against the Soviets and their proxies. Key *NSC-68* tenets included conducting offensive operations to destroy Soviet military capabilities and keep them off-balance until the full strength of the United States and its allies could be unleashed; defending the Western Hemisphere and critical allied areas to develop their war-making capacities; and aiding allies so they could carry out their tasks.[1]

Nuclear Doctrine

A particularly important factor in the development of early postwar U.S. military doctrine was the unwillingness of the United States and its North Atlantic Treaty Organization (NATO) allies to expend the resources necessary to equal the conventional force superiority of the Soviet Union and its Warsaw Pact allies. This decision required the United States and NATO to rely on the emerging nuclear weapons deterrent as the best way to preserve European peace.[2]

Consequently, one of the most critical sources of U.S. military doctrine strategy was developing documentation of the United States' willingness to use its nuclear arsenal to deter the Soviets and, if peaceful deterrence failed, to defeat them by using such weapons in war. One of the most important demonstrations of this willingness to use nuclear weapons was the Strategic Integrated Operational Plan (SIOP) issued in 1960. SIOP called for integrating the capabilities of the three nuclear weapons delivery components, or triad, which consisted of land-based intercontinental ballistic missiles (ICBMs), aerial bombers with intercontinental range, and submarine-launched ballistic missiles (SLBMs). SIOP preparation involved participation by the Joint Chiefs of Staff (JCS), the Secretary of Defense, and the President, and it detailed highly classified information on specific enemy targets the U.S. military would strike with nuclear weapons in the event of a war with the Soviet Union, China, or some other country. SIOP has been a controversial program and revising and updating it has been an ongoing process, with revisions occurring in 1962, 1976, 1981, and 1989.[3]

Massive Retaliation was another key element in early U.S. and NATO nuclear doctrinal strategy. Massive Retaliation involved NATO publicly announcing that it would respond to a Soviet bloc attack with a disproportionate response, emphasizing strategic nuclear weapons in the belief that such a policy would deter potential adversaries from initiating an attack. Another key characteristic of Massive

Retaliation was that the state that announced such a tactic had the ability to launch a second round of nuclear strikes against its attacker. Massive Retaliation was announced by Secretary of State John Foster Dulles on January 12, 1954, and it remained in force throughout the Eisenhower Administration as part of its New Look policy, which emphasized nuclear deterrence over conventional forces as the foundation of U.S. national security strategy. However, its lack of flexibility in responding to potential Soviet attack severely limited its effectiveness and it would be replaced in the Kennedy Administration.[4]

This lack of flexibility in Massive Retaliation would lead U.S. civilian and military policymakers to look for alternative responses to Soviet military attacks. The alternative decided upon was Flexible Response, which involved a mixture of conventional military force and theater nuclear forces as the bulwark of U.S. and NATO military strategy. Enunciated by Kennedy's Defense Secretary Robert McNamara, Flexible Response allowed the use of conventional defenses to stop a Soviet assault; deliberate escalation to tactical nuclear weapons if conventional defense collapsed; and escalation to strategic nuclear forces if further battlefield deterioration occurred, resulting in assured destruction of both sides.

Additional Flexible Response components included the expansion of nuclear triad development; the development of a doctrine to fight two and a half wars—with two of these conflicts being conventional wars using traditional military powers and the remainder involving fighting a brushfire conflict against irregular military forces, such as rebel guerillas; and placing key emphasis on the assured destruction aspect of a second strike by ensuring that the Soviets and other enemies understood that enough of the U.S. nuclear force could survive a first strike attack to retaliate by destroying enemy cities and industrial capacity. The doctrinal tenet of Mutual Assured Destruction (MAD) was further codified into U.S. nuclear doctrine as part of Flexible Response. Flexible Response has gone through significant evolutions since its introduction, but it has remained a critical component of U.S. nuclear doctrine until the present.[5]

The 1970s saw the previously dominant U.S. nuclear arsenal diminished by a steadily increasing Soviet nuclear capability, which caused many U.S. national security policymakers to question some Flexible Response tenets. One of these policymakers was James Schlesinger, who became President Nixon's Secretary of Defense in 1973. Recognizing that the United States no longer enjoyed nuclear superiority over the Soviets and that the Soviets now possessed an invulnerable second-strike force, Schlesinger realized that U.S. enemies would not see MAD as employable. He urged the United States to obtain more selective targeting options that were less likely to involve major mass destruction; maintain a capability to deter an enemy's desire to inflict mass destruction on the United States and its allies; and reduce U.S. targeting to enemy military targets in order to reduce potential counterattacks against U.S. cities.[6]

This new U.S. nuclear strategy, known as the Schlesinger Doctrine, was articulated on January 17, 1974 in National Security Council Decision Memorandum (NSDM) 242. Key *NSDM 242* elements included the U.S. National Command

Authority (NCA) having multiple nuclear weapons use choices and the option to escalate; an explicit U.S. targeting policy focused on selective retaliation against the enemy's military or targeted counterforce; and withholding strikes against some enemy targets and target classes so that opponents had a rational reason to terminate conflict. The Schlesinger Doctrine also sought to hold survivable nuclear forces in reserve to protect and coerce after a major nuclear conflict; destroy an enemy's critical political, economic, and military resources in order to limit an enemy's conflict recovery ability; and limit damage to critical U.S. and allied political, economic, and military resources. This doctrine also sought to ensure that NCA refined its crisis management procedures so that timely political-military assessments and recommendations concerning nuclear deployment decisions could be made to the President.[7]

The next significant document regarding U.S. nuclear weapons policy doctrine was Presidential Directive (PD) 59. This directive was issued by the Carter Administration in 1980 and stressed the Schlesinger Doctrine's counterforce modus operandi. Reflecting the work of Secretary of Defense Harold Brown, *PD 59* emphasized continuing the policy of focusing U.S. nuclear targeting on enemy military targets instead of enemy cities as a means of enhancing U.S. nuclear deterrence quality. Declassified portions of this directive stressed that U.S. strategic nuclear forces needed to be able to deter attacks against the United States and its domestic and overseas-based military forces, as well as attacks against allied countries and forces, and to deter non-nuclear attacks while targeting Soviet military and political assets, such as hardened missile and leadership relocation sites. *PD 59* went on to emphasize the United States' desire to bargain effectively to terminate a war with the most favorable terms, prevent an enemy from achieving its war aims, effectively deploy U.S. nuclear forces to work with conventional forces, and enhance the quality of U.S. command, control, communications, and intelligence capabilities.[8]

The Reagan Administration saw the first significant questioning of MAD as a U.S. nuclear doctrinal tenet. This questioning would lead to the 1983 unveiling of the Strategic Defense Initiative (SDI), which committed the United States to developing a space-based ballistic missile defense system to protect the United States and its allies from ICBM attacks. Although SDI and the idea of ballistic missile defense remain controversial, they have become an important part of U.S. nuclear doctrine by stressing the critical importance of developing effective defenses against nuclear or other weapons of mass destruction missile attacks against the United States and its allies.[9]

SDI reflected Reagan's displeasure with the inflexibility of MAD as a viable and moral position from which to defend U.S. security. Reagan and his administration believed that engaging in strenuous economic, political, and military competition with the Soviet Union would expose the weaknesses of the Communist system and expedite that system's collapse. This increased competition would see the United States increase its defense spending on both conventional and nuclear force capabilities, which would strain Soviet economic and technological capabilities and ultimately compel the Soviets to agree to nuclear arms reductions, producing a somewhat more open political system.[10]

Reagan also sought to increase pressure on the Soviets by providing military assistance to forces fighting the Soviets or Soviet-backed regimes in locales as diverse as Afghanistan, Angola, Grenada, and Nicaragua. These collective efforts became known as the Reagan Doctrine, and they would eventually succeed in compelling the Soviets to withdraw from Afghanistan, as well as achieving some domestic political reform in the Soviet Union under Mikhail Gorbachev, reaching nuclear arms control agreements like the 1987 Intermediate Nuclear Forces (INF) Treaty, retaining SDI despite Soviet attempts to eliminate the program, and beginning to move U.S. nuclear doctrine from MAD to a more flexible stance that incorporated ballistic missile defense. These developments would all play a role in the subsequent collapse of the Soviet Union and the end of the Cold War on desirable terms for the United States and its allies.[11]

The Cold War's termination reduced some tension with the former Soviet Union by reducing the size of the Russian nuclear arsenal. However, the emerging international order, as demonstrated by the 1990–1991 Persian Gulf War, saw increased emphasis on the dangers of nuclear proliferation by regimes as diverse as India, Iran, Iraq, North Korea, and Pakistan, and this was reflected in the decision-making process of national security policymakers in the George H. W. Bush Administration. This concern was reflected in *National Security Directive (NSD) 70,* issued July 10, 1992. *NSD 70* presented the tenets of U.S. nuclear nonproliferation policy by declaring its emphasis on the following:

- Total multilateral support of nonproliferation export controls, including the establishment of common enforcement standards by licensing and customs authorities;
- U.S. nonproliferation efforts focusing on areas of concern, such as the Middle East, the Persian Gulf, South Asia, and the Korean Peninsula, along with the former Soviet Union and Eastern European states;
- U.S. nonproliferation policy would seek the broadest possible multilateral support and work with organizations such as the United Nations Security Council, Enhanced Proliferation Control Initiative, and Nuclear Suppliers Group;
- The United States would examine all motivations and security rationales leading to mass destruction weapons proliferation and develop a comprehensive package of diplomatic, economic, intelligence, military, and political options to advance U.S. nonproliferation goals.[12]

The perception of a more stable international security environment with the collapse of the Soviet Union and end of the Cold War led U.S. leaders to evaluate whether nuclear weapons should continue to be tested to retain U.S. nuclear deterrent viability. On October 2, 1992, President Bush announced that the U.S. was beginning a unilateral moratorium on nuclear weapons testing. President Bill Clinton extended this moratorium in July 1993 and again in March 1994. On August 11, 1995, Clinton announced that the United States would negotiate a comprehensive nuclear test ban treaty and that it would continue its nuclear weapons testing moratorium. The United States signed the Comprehensive Test Ban Treaty on September 24, 1996, but this treaty was never ratified by the U.S. Senate.[13]

The United States sought to maintain the reliability of its nuclear weapons deterrent without conducting tests through the Stockpile Stewardship Program (SSP). This program relies on computer simulations and modeling to assess the operational viability and safety of the U.S. deterrent arsenal. This program has been in place for nearly two decades but its overall effectiveness has been questioned.[14]

Nuclear Posture Review (NPR) documents are important sources for examining recent presidential administration nuclear weapons doctrinal philosophy. The most recent versions of these documents were released in 1994 and 2001, with a related document issued in September 2008. Declassified portions of the 1994 document reaffirmed the legitimacy of the U.S. strategic nuclear deterrent and the maintenance of the existing triad of bombers, submarines, and land-based ballistic missiles. It went on to reaffirm U.S. commitments to international and bilateral arms control agreements, such as the Nuclear Nonproliferation Treaty and the Cooperative Threat Reduction Program. An additional noteworthy characteristic of this document was its call to create a "hedge force" in which warheads removed from missiles could be kept in storage to be reloaded if U.S.-Russian relations worsened.[15]

Released in January 2002, the 2001 NPR maintained the 1994 document's emphasis on combating proliferation, but also established a new triad consisting of offensive nuclear and non-nuclear strike systems, active and passive defenses, and an enhanced defense infrastructure that would provide new capabilities to meet emerging threats in a timely fashion, assisted by enhanced command and control and intelligence systems. This review also mentioned that by emphasizing defensive capabilities, the United States would no longer be as dependent on offensive strike forces for deterrence, as had been required by the Cold War, and that this deterrence enforcing capability would be bolstered by the augmented presence of conventional strike and information operations capabilities.

Additional emphases of the U.S. nuclear posture included the reduction of nuclear weapons to an arsenal of 1,700–2,200 warheads, the adjustment of U.S. strategic forces from a Cold War and Russian threat-based model to a capabilities-based approach, the continued credible deterrence of U.S. defense capabilities to nations or terrorist groups with access to mass destruction weapons and effective weapons delivery platforms, and the enhancement of U.S. defense infrastructure in order to lessen the two-decade or longer period currently required to develop and deploy new-generation weapons systems.[16]

September 2008 saw the release of the collaborative Defense Department and Energy Department report, *National Security and Nuclear Weapons in the 21st Century*. This document noted developments in the Chinese and Russian nuclear arsenals, mentioning that the Russians maintained a fully functioning nuclear weapons design, development, test, and manufacturing infrastructure capable of producing significant numbers of nuclear warheads per year and that increased emphasis had been placed on nuclear weapons in Russian national security policy and military doctrine, in addition to reincorporating theater nuclear options into its military planning.[17]

This report further noted recent changes in British and French nuclear weapons capabilities and stated that the focus of U.S. military deterrence included assuring its friends and allies, dissuading nations from military competition with the United States, deterring adversaries from attacking the United States, and defeating such attacks if necessary. It also noted that SIOP was replaced in 2003 with a plan that provided more flexible targeting options and that the United States was on its way to meeting 2001 NPR goals of reducing its operationally deployed nuclear warheads to a total of 1,700–2,200, which would be composed in 2012 of a mixture of Minutemen ICBMs, Ohio class submarine ballistic missiles, and B-2 and B-52 bombers.[18]

National Security and Nuclear Weapons in the 21st Century also maintained that SSP had been successful and that the United States' nuclear warhead arsenal was safe, secure, and reliable. It acknowledged, however, that current strategies may be unsustainable in the future and that national nuclear weapons laboratory directors had expressed concern about ensuring confidence in the legacy stockpile's long-term reliability without nuclear testing. The Reliable Replacement Warhead (RRW) program has been proposed as a means of revitalizing the United States' nuclear arsenal by producing new warheads to meet future requirements for maintaining the quality and reliability of U.S. nuclear weapons. RRW would also make it possible to improve weapon security features to prevent their accidental and unauthorized use and to reduce the possibility of needing to conduct underground nuclear weapons tests to certify weapon reliability.[19]

Emerging U.S. nuclear weapons doctrine will focus on preventing nuclear proliferation to countries of concern, such as Iran and North Korea, and to terrorist groups, continuing prudent reductions in the U.S. nuclear weapons arsenal while allowing for agile responses to potential threats, and seeking to develop ways to ensure the reliability of this arsenal without resuming underground weapons testing. Review of these matters will require continued involvement by presidential administrations, the Energy and Defense Departments, the military, and congressional oversight committees.

Air Force Doctrine

U.S. Air Force doctrinal history has been complicated and subject to often considerable criticism for neglecting airpower theory, which one critic contends has impaired its ability to write sound doctrine, including operational doctrine. This critic goes on to maintain that the Air Force needs an established and institutionalized process for developing and transmitting basic and operational level doctrine, that the service fears it will doctrinally commit itself to more than it can deliver, and that a paranoid mentality about service survival has caused the Air Force to emphasize winning budget battles for equipment instead of developing an all-encompassing airpower theoretical foundation.[20]

The multivolume *United States Strategic Bombing Survey,* which documented the results of U.S. aerial bombing of Germany and Japan during World War II, was an important example of an emerging Air Force doctrinal advocacy of the wartime efficacy of aerial bombing.[21] When the Air Force achieved independence from the Army on September 18, 1947, early Air Force leaders like General Carl Spaatz (1891–1974) helped create a force structure with a Tactical Air Command, which influenced future Air Force doctrine along with the Air Force's professional military educational institution, Air University at Maxwell Air Force Base, AL.[22]

United States Air Force Basic Doctrine AFM 1–2, issued in March 1953, proved to be the Air Force's first authoritative doctrinal publication. This document saw Air Force Chief of Staff General Hoyt Vandenberg (1899–1954) provide the following perspective on airpower doctrine:

> basic air doctrine evolves from experience gained in war and from analysis of the continuing impact of new weapon systems on warfare. The dynamic and constant changes in new weapons makes periodic substantive review of this doctrine necessary.[23]

1953 and 1954 saw the release of additional Air Force doctrine publications covering subjects such as theater air operations, air defense operations, aerial and amphibious operational collaboration, and strategic air operations.[24]

The 1950s would also see various technological developments that would pose acute challenges to nascent Air Force doctrinal perceptions limited to aerial combat. The emerging Soviet nuclear ballistic missile arsenal and the Sputnik satellite launch forced the Air Force to recognize the increasing importance of space in military affairs, forcing it to extend the conception of its mission responsibilities to space and to combine air and space operational activity with the term aerospace, which would become an important and continually debated area of Air Force doctrinal mission emphasis.[25]

In March 1963, Air Force Chief of Staff Curtis Lemay (1906–1990) initiated Project Forecast—a comprehensive examination of technology and the role it might play in Air Force operations. This program would identify potential Air Force opportunities in technical areas such as materials, propulsion, flight dynamics, guidance, and computer technology that could benefit service operations and help develop Air Force technology into the 1980s. While Forecast was ongoing, Air Force Secretary Eugene Zuckert (1911–2000) sought to change its conceptual doctrine approach. Zuckert believed Air Force doctrine needed to be written to support national policy and strategy as opposed to being an airpower theory based on aerospace doctrine, which was rooted in operational experience and which reflected the peace and wartime capabilities and limitations of aerospace forces. Lemay and Zuckert's collaborative efforts would result in the August 1964 issuance, *United States Air Force Basic Doctrine AFM 1-1,* which maintained that basic doctrine evolves through ongoing military operations testing and analysis in light of existing national objectives and changing military environments.[26]

The follow-up manuals that implemented *AFM 1–1* would serve as the basis for the burgeoning U.S. aerial military involvement in the Vietnam War. These manuals included *AFM 2–1 Tactical Air Operations-Counter Air, Close Air Support, and Air Interdiction; AFM 2–3 Air Operations in Conjunction with Amphibious Operations;* and *AFM 2–4 Assault Airlift. AFM 2–1* introduced the idea of sortie apportionment and addressed aerial interdiction to give operators an idea of how to plan such efforts. It also included chapters on close air support and the order in which theater forces should accomplish specific objectives. One critic of this work says it also reflected the Air Force's reluctance to specify what it could really accomplish in war because it was institutionally fearful of promising more than it could deliver.[27]

The Vietnam War had a profound impact on Air Force doctrine. It illustrated the consequences of the United States' fixation on nuclear strategy at the price of sufficient preparations for conventional war—let alone airpower in counterinsurgency warfare. Vietnam also demonstrated the consequences of having unclear policy goals and committing airpower haphazardly instead of with determined resolve. Additionally, the war demonstrated ongoing problems with conducting modern conventional war against a well-equipped and sophisticated opponent, while also providing a clear indicator of the defenses the United States and its NATO allies would have to deal with if counteroffensive operations against Soviet and Warsaw Pact forces were conducted in Central Europe.[28]

The next edition of *AFM 1–1* was released in 1984. Reflecting the influence of Soviet military strategy, the experiences of the 1973 Arab-Israeli war, and a growing belief within the military for the need for greater force integration and collaboration among service branches, this document discussed many of the military doctrine aspects reflected in Reagan Administration defense policy planning. This document, entitled *Basic Aerospace Doctrine of the United States Air Force,* included the chapter, "Employing Aerospace Forces," which discussed man, machine, and the environment as interacting war-fighting principles and listed economy of force, maneuver, timing and tempo, command unity, simplicity, logistics, and cohesion as hallmark military principles. This new doctrine stressed that aerospace forces must work with land and naval forces in unified action, that the Air Force must contribute to the success of maritime missions, and that many Air Force missions could be performed from space-based platforms. *AFM 1–1* also continued service adherence to traditional doctrinal precepts such as air superiority being a first consideration in employing aerospace forces, airpower's ability to exploit speed, range, and flexibility more quickly than land and sea forces, and that speed, range, and flexibility could be best utilized when airpower was centrally controlled and de-centrally executed.[29]

The Persian Gulf War of 1990–1991 was an excellent example of the Air Force demonstrating its technological and operational superiority over Iraqi forces in relatively swift and low-cost operations. U.S. air operations in this conflict demonstrated the value of precision-guided munitions, day-night all-weather operations, and space-based assets, such as global positioning satellites, in destroying or disabling enemy military assets.[30] This conflict gave the Air Force the opportunity to conduct extensive assessments of its operations and the relevance of Air Force doctrine to current and future combat operations. The collective result of these efforts was the 1993 *Gulf War Airpower Survey,* a five-volume after-action assessment of Operation Desert Storm. The first volume discussed coalition plans to achieve aerial superiority and analyzed command and control issues essential to effective airpower usage. The second volume discussed how airpower was used to destroy Iraqi military forces and examine coalition airpower operational level accomplishments. The third volume discussed air operation logistics, the fourth volume scrutinized the role of weapons, tactics, and training in Gulf War airpower employment and force projection and provided a brief, unclassified summary of space operations, and the fifth volume featured a statistical compendium of aerial operations and a chronology of key events during this conflict.[31]

The aftermath of the Gulf War and the successful role played by space assets in Air Force operations prompted an attempt to incorporate space into service doctrine in the March 1992 edition of *AFM 1–1*. The first volume of this manual stressed that Air Force doctrine emphasized the nature of aerospace power and the operational art of employing and preparing aerospace forces for war. The second volume sought to provide factual support for Air Force basic doctrine, including the importance of educating, equipping, training, and organizing the Air Force to meet its responsibilities.[32]

Additional attempts to integrate space into Air Force doctrine include the *Spacecast 2020* and *Air Force 2025* studies of 1994 and 1995–1996. *Spacecast 2020* emphasized the critical roles played by space transportation and the U.S. commercial space launch industry in ensuring the development and maintenance of space-based lasers that feature surveillance and counterforce capability and space-lift vehicles, as well as the importance of integrating space doctrine into professional military education.[33]

Air Force 2025 discussed how space doctrine and strategy might be integrated into future aerospace military operations. Topics addressed in this multivolume compilation include the importance of space lift to space superiority, the criticality of vertically integrated planning, the development of smarter technological procurement methods, the importance of integrating information operations into aerospace war-fighting doctrine, methods to effectively incorporate interdiction into such operations, and how weather control can be a critical factor in determining the success or failure of aerospace operations.[34]

NATO's 1999 military operations against Serbia in Operation Allied Force would also raise questions about airpower's doctrinal and operational feasibility in a military campaign. This operation succeeded in causing Serbian leader Slobodan Milosevic's (1941–2006) regime to give up its efforts to retain Kosovo as part of Serbia. However, there was and is ongoing debate over whether airpower alone can achieve desired military objectives or whether it must be combined with land power. The decision of U.S. and NATO leaders to rule out a ground invasion of Serbia allowed Serbian atrocities against Kosovar Albanians to continue while the aerial campaign was waged. Further, it was an ineffective use of airpower in a major military operation because it failed to achieve surprise and keep the Serbs unaware of NATO military intentions.[35]

The most recent U.S. Air Force Doctrine Document (AFDD) is *Air Force Basic Doctrine 1* (*AFDD-1*), issued in 2003. Described in greater detail in the next chapter, its contents cover topics such as the nature of air force doctrine, strategic attack, expeditionary air force organization, and attributes such as global attack and precision engagement.[36] Current *AFDDs* cover topics such as air warfare, strategic attack, nuclear operations, airspace control in combat zones, and other topics reflective of the Air Force's multifaceted missions.[37]

U.S. Air Force doctrine has experienced considerable evolution over its six-decade history. This literature encompasses the conduct of conventional and nuclear force operations, the use and increasing importance of space in military affairs, and how to optimize aerospace power in conducting counterinsurgency operations against terrorist groups or other countries. There is ongoing debate over the future direction and viability of U.S. Air Force doctrine. This debate concerns the role, if any, that military space operations should play in U.S. military doctrine and the role of the Air Force in future U.S. military operations. Those who wish to study Air Force doctrine have access to ample resources from participants in this debate, as is demonstrated in later sections of this book.

Army Doctrine

U.S. Army doctrine has experienced revolutionary and evolutionary changes in the six decades since World War II. Topics addressed by Army doctrine during this time period include the conduct of U.S. and NATO operations in conventional and nuclear environments; approaches to conventional force operations involving infantry, artillery, and armored forces; coordination of operations with other U.S. and allied armed services; peace support operations; humanitarian operations; counterinsurgency operations, which have assumed preeminence in a post–9/11 world; and the legal and normative implications of enemy combatant detainees. A variety of sources have sought to document how the Army has responded to these doctrinal issues.[38]

The Pentomic Army concept was a significant proposal to develop an Army force capable of fighting the Soviet bloc forces it expected to face in European combat after World War II. This structure was adopted by the Army in 1957 in response to the threat of tactical nuclear weapons to battlefield force structure. Under this Pentomic structure, an Army division was organized into five battle groups, each commanded by a colonel. These battle groups had five rifle companies, a combat support company, and a headquarters company commanded by a captain. Artillery units in this structure were organized in five batteries with four of these being howitzer batteries and the fifth a mortar battery. The Pentomic Army sought to further the Eisenhower Administration's New Look policies by emphasizing reliance on nuclear weapons and featuring nuclear-capable rocket and tube artillery, high mobility, and effective communication. This structure did not improve Army fighting capability, but it did help stabilize the declining funding and staffing structure that threatened Army operational capabilities during this period.[39]

The Pentomic Army would be replaced in 1961 by the Kennedy Administration's Reorganization Objective Army Division (ROAD) as part of the United States' shift to flexible response as its nuclear deterrent doctrinal strategy.

ROAD characteristics included being able to operate in nuclear and nonnuclear environments and being able to add a flexible number of maneuver battalions to increase armor or infantry strength as battlefield situations permitted. ROAD divisions would become standardized, featuring armored divisions with six tank and five mechanized infantry battalions; mechanized divisions with three tank and seven mechanized infantry battalions; infantry divisions with two tank and eight infantry battalions; and airborne divisions with one assault gun battalion and eight airborne infantry battalions. ROAD was accepted by Defense Secretary Robert McNamara in 1961, but it was not fully integrated into the United States' European Army until 1963, and its order of battle by mid-1964 was 22 tank battalions, 31 mechanized battalions, three infantry battalions, zero airborne battalions, and 56 maneuver battalions.[40]

ROAD remained the overall organizational structure for the Army's European forces, but the Vietnam War required increasing numbers of troops and many of these were transferred to Vietnam, which drastically reduced the quality of U.S. forces in Europe and made U.S. relations with its NATO European allies more difficult.[41] Political and military controversy over the Vietnam War would result in the United States being forced to withdraw, producing a 1975 Vietnamese Communist triumph. The U.S. Army largely tried to fight this war with conventional military doctrine instead of counterinsurgency doctrine, and its failure to adapt to counterinsurgency combat environment requirements was one of the many reasons for the traumatic U.S. defeat. This defeat would cause the Army and other U.S. military service branches to engage in extensive critical analysis of the reasons for this defeat, the lessons which could be learned from this failure, and the possible ways these lessons could be applied to future military conflicts the United States might face.[42]

An important example of this post–Vietnam Army retrospection was the decision to establish a Training and Doctrine Command (TRADOC) in 1973.[43]

TRADOC was created to centralize and coordinate Army training and doctrine programs. Generals William DePuy (1919–1992) and Donn Starry were prominent early TRADOC leaders who helped sculpt the Army's attempts to develop post–Vietnam doctrine and analyze the lessons learned from that conflict.[44]

DePuy was convinced that the ending of the draft in 1973[45] would leave the Army with a limited recruiting pool and he was especially concerned about small-unit leadership quality without the draft. He believed the Army should be rebuilt by designing tactical and operational doctrine for a full-scale Warsaw Pact offensive in Germany instead of coping with counterinsurgency conflicts such as

Vietnam. DePuy believed TRADOC should design doctrine first and that command structures including equipment requirements should be customized to match such doctrine. He was convinced that the tank remained the critical weapon in the Central European military environment and that tank defensive capabilities and new antitank missiles favored outnumbered defenders fighting Soviet bloc forces.[46]

Working with the German Bundeswehr in NATO, DePuy also became concerned about the compatibility of German and American operational concepts and came to believe that German tactics regulating the close cooperation of tanks and armored infantry to be superior to comparable U.S. Army tactics. DePuy believed that the opening battles of the next European war would be fought on the defensive with armored and mechanized forces augmented by wire-guided antitank missiles, and he preferred the German concept of putting the preponderance of active forces at the front of the battle zone to facilitate their response to an invading force.[47]

TRADOC leaders were also profoundly influenced by the October 1973 Arab-Israeli War. This conflict demonstrated that contemporary battlefields could produce considerable destruction in a short time and that the U.S. military would no longer have the lengthy time frames it had traditionally had to mobilize its forces before sending them into battle. Consequently, it would be incumbent upon the Army to have prepositioned equipment and trained forces ready to be sent into combat environments. Investment in new technology would also be required considering the more lethal battlefield produced by the Yom Kippur War.[48]

A visible manifestation of TRADOC-produced doctrinal thinking was the 1976 edition of *Field Manual (FM) 100–5* as the Army's principal fighting document. *FM 100–5* emphasized that the Army must prepare to win the first battle of the next war and all subsequent battles. It went on to stress that the Army needed a clear, coherent, and rigorous doctrine capable of ensuring each of its weapon systems was deployed with optimum effectiveness. A key doctrinal assumption of *FM 100–5* was that war could begin conventionally, move into a combined conventional-nuclear phase, and return to a conventional battle. *FM 100–5* also stressed the use of nuclear weapons against second echelon or reserve forces and that tactical advantage could be gained by neutralizing lead enemy second echelon elements by eliminating this echelon's support and supporting fire systems while destroying follow-up reserves and reducing pressure on allied forces so they could contain engaged forces by conventional methods and control the battlefield.[49]

An additional *FM 100–5* characteristic was its stress on changes in military operations caused by mobility advances, night-fighting capabilities, electronic warfare, and a growing emphasis on air-land operational mobility. It also emphasized better training, suppressive tactics, effective terrain use, and combined arms coordination to counter increased weapons lethality.[50] Despite these areas of emphasis, *FM 100–5* received considerable criticism within sections of the Army doctrinal community. Considerable criticism of *FM100–5*'s active defense provisions was expressed in the Army Command and General Staff College's journal, *Military*

Review, including the contention that active defense might be able to defeat the initial Soviet assault but that U.S. and allied forces would be overrun by follow-up Soviet bloc forces and that there was insufficient Army organizational consensus behind *FM 100–5* precepts.[51]

Efforts to reach an Army doctrinal organizational war-fighting consensus occurred under Donn Starry, who succeeded DePuy as TRADOC commander in 1977. Starry would set in motion revisions to *FM 100–5* that would produce the doctrinal concept of AirLand Battle. A contributing factor to Starry's AirLand Battle promotion was his belief that the military axiom that an attacker should have at least a three-to-one force ratio over the defender was flawed. He believed that historical tank battles demonstrated that there was little difference in battle outcomes as long as the attacker did not have at least a six-to-one force ratio superiority over the defender.[52]

The late 1970s and early 1980s would see Starry and other Army doctrinal planners develop a plan to integrate armor, mobile infantry, artillery, missile forces, and airpower. Starry believed that battlefield developments could be statistically determined in areas such as minutes into battle, force ratios, specific weapons, rates of advance, visibility, rate of fire, number of command decisions, and time from request to tactical air support delivery. He believed that an attacker needed a better than five-to-one numerical advantage to defeat prepared and determined defensive forces. AirLand Battle had an explicitly offensive emphasis and sought to provide an extended chronological, territorial, and spatial view of the battlefield.[53]

AirLand Battle was published in August 1982 in an updated edition of *FM 100–5* and incorporated German operational concepts such as *Auftragstaktik* and *Schwerpunkt* into its modus operandi. *Auftragstaktik,* or mission order tactics, allows for greater decision-making by tactical-level commanders, and *Schwerpunkt* refers to center of gravity where forces and assets can be shifted to achieve breakthroughs against enemy forces.[54] This work provided a detailed scenario for a second-echelon attack against enemy forces beginning with battlefield intelligence preparation in which commanders, aided by a sophisticated sensor and communications systems network, would attack high-value targets to disrupt enemy forward momentum. Such attacks would occur through interdiction (including airpower, artillery, and special forces), offensive electronic warfare, and deception. AirLand Battle stressed the critical imperative of an integrated attack plan aimed at enemy assault and follow-on forces, with airpower dominating the early phases of this battle. Particular emphasis was placed on avoiding the enemy's main strength, and shattering its will by reducing its fighting capability was represented as the fastest and cheapest method to win wars.[55]

AirLand Battle would receive its penultimate testing and demonstration in Operation Desert Storm against Iraqi forces in 1990–1991. Instead of fighting Soviet forces, U.S. and coalition forces fought and easily defeated Soviet-trained and -equipped Iraqi forces using AirLand Battle doctrinal precepts that included the successful integration of aerial and ground forces and the superior initiative and training of coalition forces.[56]

A related adjunct concept to AirLand Battle adopted by U.S. and NATO forces in the late 1980s would be Follow on Forces Attack (FOFA). FOFA involved the use of various conventionally armed long-range weapons to attack Warsaw Pact ground forces that had not yet engaged NATO forces. Its purpose was to delay, disrupt, and destroy these follow-on forces so that NATO defenses could hold as far forward as possible in the Central European battlefield, emphasizing the area where West Germany bordered East Germany and Czechoslovakia. There was controversy over FOFA in some NATO countries because it required an increase in defense budgets beyond the three percent real growth to which they had committed in the 1980s. NATO's ability to fully implement FOFA was also limited by insufficient resources for reconnaissance, surveillance, and targeting acquisition, insufficiently capable munitions and weapons with which to distribute these munitions, and total systems from surveillance to target destruction capable of responding rapidly, effectively, and flexibly across large geographic areas.[57]

The end of the Cold War, the collapse of the Soviet Union and the Warsaw Pact, and the U.S. victory in the Persian Gulf War would produce reductions in U.S. defense spending and force size and a search for a new national security strategy. The Army was affected by these upheavals but soon found itself playing an increasingly important role in conducting peacekeeping and stabilization operations in the early post–Cold War era, and recognition of this changing reality occurred in the updated June 1993 edition of *FM 100–5*.[58] This changing Army operational combat role occurred most vividly in Somalia, where the United States sought to stabilize security conditions involving warring factions in that conflict-ridden county only to be caught in a nasty civil war where U.S. forces became combatants and suffered fatalities before being withdrawn by the Clinton Administration. U.S. Army forces also became involved in peacekeeping operations in Bosnia and Kosovo, where they sought, with more success than in Somalia, to stabilize conditions after the violent conflicts that followed Yugoslavia's disintegration. Participation in these peacekeeping operations, whether done in concert with NATO or United Nations mandates or in cooperation with host countries, would stimulate considerable debate and controversy within the U.S. military as it sought to develop a coherent and sustainable military doctrine for conducting such operations that ran counter to Army military doctrine, which traditionally emphasized victorious conventional war-fighting.[59]

The relative calm of U.S. Army military doctrine development during the 1990s would be shattered by the 9/11 al Qaida terrorist attacks against New York City and Washington, DC and the subsequent and ongoing U.S. military operations in Afghanistan and Iraq. These operations forced the U.S. military to rediscover the importance of counterinsurgency (COIN) operations, which had been deemphasized after reviewing the lessons learned in Vietnam.[60] The evolving domestic and international political, diplomatic, and economic realities of these conflicts have also affected U.S. army doctrinal writers as they have sought to build the political, socio-economic, and military infrastructures necessary to help governments and tribal groupings in these countries create the political stability essential to defeat

Al Qaida and Taliban forces and to build nation-states capable of standing on their own and resisting Islamist terror.[61]

This reassertion of counterinsurgency's importance in Army doctrine has been incorporated into *FM 1 The Army: Our Army at War: Relevant and Ready Today and Tomorrow.* This document, which was released in June 2005 as the Army's strategic doctrinal keystone publication, states that threats to U.S. interests may come from traditional sources, such as nation-states, and nontraditional sources, such as terrorist groups that may use unconventional methods and weapons of mass destruction. *FM 1* also maintains that non-state threats may be loosely organized networks or cells that are based on beliefs and criminal activities instead of hierarchical structures; such cells possess minimal physical presence, are difficult to target, and have no moral obligation to limit collateral damage. These threats are also elusive and seek to conceal themselves in complex natural or human geographic environments, which makes it difficult to acquire the accurate and comprehensive intelligence necessary for effective precision attacks against them and which limits Army commanders' flexibility to freely determine the time and place of engagement.[62]

The increased emphasis on asymmetric, as opposed to conventional, conflict as a focus of Army operational planning is also reflected in the following passage:

> In the aftermath of 11 September 2001, it is inadequate to focus defenses only on threats by other states and known enemies. The strategic environment requires the Army to respond to unconventional and asymmetric threats too. The most prominent are followers of extremist ideologies. The protection afforded by geographic distance has decreased, while the potential for attacks on civilian, military, and economic targets has increased. The threat of an attack with weapons of mass destruction or other means of causing catastrophic effects adds urgency to operations against these enemies. The current trend toward regional and global integration may render interstate war less likely. However, the stability and legitimacy of the conventional political order in regions vital to the United States are increasingly under pressure.[63]

The current emphasis, if not preeminence, on developing a doctrinal response to counterinsurgency warfare was reflected in the December 2006 update of *FM 3–24 Counterinsurgency* as the Army's doctrinal guide for conducting counterinsurgency operations. The writing of this joint Army and Marine Corps publication shows the heavy influence of General David Petraeus and stresses topics such as integrating civilian and military activities, the criticality of intelligence in battlefield planning and preparation, source protection, developing host nation security forces, maintaining ethical conduct toward indigenous inhabitants, distinguishing between war-fighting and policing, developing effective and legal detention practices, ensuring proper U.S. force discipline, and providing humanitarian relief and reconstruction.[64]

Emerging and future U.S. Army doctrine will focus on the multiple legal, military, normative, and operational complexities involved in conducting counterinsurgency military operations in Afghanistan, Colombia, Iraq, and other global crisis areas where the United States may be required to use its military forces. Army doctrine will also continue to focus on conducting conventional military operations in areas such as Iran and North Korea, and on defending against a Chinese invasion of Taiwan. Army doctrine also will continue to address battlefield operations in nuclear or other WMD combat environments, space operations, and information warfare given the exponential technological advances that have made these areas potential military operational venues. An ample knowledge base of scholarly and popular analysis of Army doctrinal literature currently exists, and its further development will continue to prompt additional scrutiny.

Marine Corps

The genesis of modern U.S. Marine Corps doctrinal thinking begins with the 1940 publication of its *Small Wars Manual.* This work sought to compile information gleaned by the Corps from its experience conducting counterinsurgency warfare during early 20th-century campaigns in locales as varied as China, Latin America, and the Philippines. It placed significant emphasis on historical experience and divided counterinsurgency pacification campaigns into five phases: intervention, field operations, transferring control to indigenous security forces, holding elections, and withdrawing. *Small Wars Manual* has experienced ebbs and flows in usage. It does not appear to have been consulted during the Vietnam War but its contents are particularly relevant for ongoing wars in Afghanistan and Iraq.[65]

An institutional cultural challenge faced by the Corps in its effort to develop its own service doctrine stems from the recognition that, as a combined air and ground military force operating from the sea, many of its functional activities duplicate Army and Air Force missions. Recognition of this duplication is well understood by the Corps, and Marines recognize that there have been historical instances in which the Army and Air Force have sought to undermine or eliminate them as an institution. This has, in turn, led the Corps to vigilantly maintain its institutional identity and unique mission against real or perceived encroachments by other armed services, and has produced heightened sensitivity to changes in the United States' military strategic environment that might injure the Corps's sense of identity and mission.[66]

An early post–World War II doctrinal issue confronted by the Corps was the belief of some military leaders, following the Operation Crossroads atomic bombs test during Summer 1946, that sufficient damage was done to the surrounding environment to drastically alter and potentially negate the utility of World War II-style amphibious warfare, which were core components of Marine Corps and Navy mission emphases. Both services disagreed with this assessment and contended that amphibious assaults could be conducted in a nuclear environment if there was increased naval air and surface fleet dispersion and if greater use was made of helicopters in amphibious operations.[67]

The Marine Corps's ability to effectively argue its institutional requirements was strengthened by 1952 legislation that gave it an equal voice in Joint Chiefs of

Staff military policy deliberations.[68] The Korean War was raging when this statute was enacted, and it saw the Corps make the first use of helicopters to transport and supply troops to support ground operations in Operation Windmill in the Soyang River region on September 13, 1951. Korea also saw numerous doctrinal documents on airpower's integration into Marine operations, such as *General Order 85* on February 15, 1951, which announced the policy of vertical envelopment as a means of providing aerial support to combat units. Additionally, February 1953 saw the issuance of *Landing Force Bulletin (LFB) 2 Interim Document for the Conduct of Tactical Atomic Warfare*, which prescribed operational conduct when nuclear weapons were used. The 1950s also saw the issuance of *LFB 17 Concept of Future Amphibious Operations* and *LFB 24 Helicopter Operations*, which sought to detail Corps doctrine in these operational activity areas.[69]

The Kennedy Administration's emphasis on flexible response as its nuclear doctrine and the President's interest in and support for special operations forces gave new support to the Corps's interest in limited wars. The Corps received tangible benefit from flexible response when the administration recommended that its maximum force strength be increased from 170,000 to 190,000 and that its budget be increased by $67 million to pay for new personnel and expedited modernization.[70]

Like other services, the Corps played a significant role in the Vietnam War, beginning with the March 8, 1965 landing of the 9th Marine Expeditionary Brigade at Da Nang. During Vietnam, the Corps participated in civic action programs such as provincial reconstruction as well as combat operations that emphasized pacification. Helicopter operations and doctrine received increasing use and emphasis during this war, as did vertical/short-take-off and landing (VSTOL) aircraft, which were used to expand the Corps's striking power and mobility.[71]

An initial Corps post-mortem assessment of Vietnam was provided in 1971 by Marine Corps Commandant General Leonard Chapman (1913–2000), who contended that the United States had been defeated and thrown out and that the best approach was to forget about it. This amnesiac approach prevented the Corps from seriously debating Vietnam until the late 1970s and early 1980s.[72] The immediate post-Vietnam aftermath saw the Corps seek to reassert its identity as a seaborne force specializing in amphibious warfare, its role in defending Europe against a Soviet attack, and the steps the Corps should take against a Soviet European assault or against a Soviet-style assault fielded by nations in crisis areas like the Middle East.[73]

A 1976 Brookings Institution study questioned the viability of a central European front mission for the Corps in its current condition. The study argued that to have such a role in European defense, the Corps would need to transform itself into an organization like the U.S. Army, equipped for sustained inland European combat. Such a change would effectively eliminate the Corps's *raison d'être* as an amphibious force, and the Brookings study questioned the utility and feasibility of amphibious operations in light of the Soviets' emerging and abundant arsenal of long range, highly lethal, and accurate weapons.[74]

Leading Corps doctrinal planners responded to this by urging increased mechanization and armor in their organization so that they could serve as a credible European fighting force with strengthened amphibious capabilities, while also placing emphasis on having a greater role in the Asian operational theater and creating an airborne force.[75] Consequently, Corps programs in the late 1970s and early 1980s stressed the imperative of improved tactical and strategic mobility and the combined use of air and ground units. Marine doctrine sought to emphasize air and ground unit operational adhesion by stressing the Marine Air-Ground Task Force (MAGTF), which consisted of Marine Expeditionary Units (MEU), Marine Expeditionary Brigades (MEB), and Marine Expeditionary Forces (MEF) comprised of battalion landing teams, tactical air and helicopter squadrons, and combat support and service units that would facilitate the successful completion of coordinated air and ground operations.[76]

The need for mobility and flexibility in Corps missions became tragically apparent in 1983, when Islamist terrorists conducted successful suicide bombings at Marine bases in Beirut, Lebanon as part of a successful attempt to drive the U.S. military from its peacekeeping responsibilities following Lebanon's civil war.[77] This tragedy was facilitated by civilian and military government policymakers who put highly mobile forces into a dangerously unprotected and static position, local Corps commanders who failed to anticipate such an attack and protect their forces, and the military's inability to effectively process and interpret significant quantities of human intelligence that indicated the probability of such an attack. This tragedy also helped Corps leaders recognize that they needed to focus on operational tactics that would be used against the Corps and other U.S. military personnel in future combat operations in the Middle East.[78]

The 1980s would also see the augmentation of the Corps's traditional emphasis on expeditionary warfare with maneuver warfare development, as embodied in AirLand Battle. Maneuver warfare has increasingly become a preeminent focus of Marine Corps doctrine, and it has received particular emphasis in *Fleet Marine Force Manual 1 (FMFM-1) Warfighting*, published in 1989.[79] The following passage from *FMFM-1* stresses that maneuver warfare differs significantly from attrition warfare by placing greater emphasis on circumventing a problem and attacking it from a favorable position instead of meeting it head on:

> maneuver relies on speed and surprise, for without either we cannot concentrate strength against enemy weakness... The need for speed... requires decentralized control. While attrition operates principally in the physical realm of war, the results of maneuver are both physical and moral. The object of maneuver is not so much to destroy physically as it is to shatter the enemy's cohesion, organization, command, and psychological balance. Successful maneuver depends on the ability to identify and exploit enemy weakness, not simply on the expenditure of superior might. To win by maneuver we cannot substitute numbers for skill. Maneuver thus makes a greater demand on military judgment. Potential success by maneuver unlike attrition is often disproportionate to the effort made. But for exactly the same reasons, maneuver

> incompetently applied carries with it a greater chance for catastrophic failure, while attrition is inherently less risky.[80]

The 1990s saw the Marine Corps participate successfully in Operation Desert Storm and seek to develop a mission for its expeditionary and amphibious operational strengths while the United States sought to develop viable post–Cold War national security strategies in an international security environment that included terrorism and unconventional military operations in locales as diverse as the Balkans, Rwanda, and the Middle East.[81]

Recent and ongoing U.S. military operations in Afghanistan and Iraq have given the Marine Corps significant roles in these theaters and resurrected the Corps's traditional emphasis on fighting small wars using counterinsurgency doctrine. This has required the Corps and other U.S. armed services to recognize that successful antiterrorism campaigns require high levels of cultural sensitivity; a recognition of the interrelationship between political and military goals, including building and strengthening indigenous armies and police forces; and the importance of cultivating and sustaining the support of the local populations in these countries.[82]

In addition to the Afghanistan and Iraq operations, the Corps continues to stress the importance of its expeditionary warfare capabilities. The *Marine Corps Strategy 21* document issued in 2000 stresses that expeditionary maneuver warfare (EMW) is the Corps's capstone operational principle, which incorporates previously published operational concepts such as Operational Maneuver from the Sea and Ship to Objective Maneuver. EMW emphases include:

- *Joint enabling:* The ability to use Marine forces to serve as a lead element of a joint task force;
- *Strategic agility:* The ability to transition rapidly from pre-crisis readiness to full combat capability while deployed in a distant theater;
- *Operational reach:* The ability to project and sustain relevant and effective power across the depth of a battle-space;
- *Tactical flexibility:* The capability to conduct a range of dissimilar missions, *concurrently*, in support of a joint team across the entire spectrum of conflict.[83]

The 2008 Marine Corps strategic planning document stresses that the Corps will seek to implement its doctrinal objectives and mission requirements by increasing its personnel from 175,000 to 202,000 between Fiscal Years 2008–2011, and that its force modernization efforts will place particular emphasis on acquiring force protection personal protective equipment to protect against improvised explosive devices and the dangers involved in seeking to dispose of explosive ordnance in combat zones as well as the dangers of dealing with weapons of mass destruction.[84] Corps doctrine will continue to adapt to cope with the constantly changing requirements of conducting counterinsurgency warfare against agile and adaptive enemies, while seeking to update the Corps's emphasis on conducting successful amphibious and littoral expeditionary operations against nation-states, terrorist organizations, or criminal groups.

Navy

The United States Navy entered the post–World War II period having successfully defeated German and Japanese naval forces. The emerging postwar global security environment emphasized the importance of developing a nuclear deterrent to restrain Soviet bloc forces, which were numerically superior to U.S. forces and their allies. The 1947 National Security Act created a separate Air Force, but retained naval aviation as a separate fleet function and allowed the development of naval aviation despite Air Force resistance.[85]

The earliest postwar Navy strategy document was a November 5, 1945 proposal by Vice Admiral Harry Hill (1890–1971) that advocated global military containment of the Soviet Union. This proposal was formally endorsed by the Joint Chiefs of Staff (JCS) on July 27, 1946, and Naval War College studies prepared at this time pointed to the North Atlantic and Eastern Mediterranean as areas needing a significant U.S. naval presence. Hill's proposal was ultimately incorporated into an early 1947 Naval Maritime Strategy and into President Harry Truman's March 1947 announcement of Soviet Union containment as a U.S. national security strategy.[86]

An early postwar Navy doctrinal document was *Principles and Applications of Naval Warfare: United States Fleets USF-1*, issued in 1947 by Chester W. Nimitz (1885–1966). This document set forth general principles for the Navy to conduct future wars and included a chapter on cooperating with allied navies.[87] The *Naval Manual of Operational Planning* (1948), a supplemental document prepared by the Chief of Naval Operations, has served as a foundation for much modern naval doctrinal planning.[88]

This period would see the Navy fight to preserve its belief in the importance of its mission to engage in surface warfare operations, conduct amphibious operations, and retain a naval aviation program to support the extension of naval firepower into future operational theaters. Sentiment existed within significant military circles at this time that strategic aerial bombing and large-scale land operations were the prevailing military operational trends. This mindset was vividly demonstrated when Joint Chiefs of Staff Chair General Omar Bradley (1893–1981) told a congressional committee on October 19, 1949 that there was no longer any need for Pacific Ocean style island-hopping campaigns and that large-scale amphibious operations such as those occurring in Normandy and Sicily during World War II would never happen again.[89]

The Navy would spend significant time fighting the Air Force and the Army in the immediate postwar period to retain its aviation assets. These were considered necessary if the Navy was to use aircraft carriers and planes launched from those carriers to conduct conventional and nuclear strikes against targets in the Soviet Union or China. The Navy was successful in sustaining its aviation program thanks to successfully cultivating congressional support in testimony during October 1949 hearings, and its differences with other services were temporarily submerged by the 1950 outbreak of the Korean War.[90]

This conflict would see the Navy play a particularly important role in the successful September 1950 amphibious invasion of Inchon, which helped turn the war in favor of the allies until Chinese intervention later that year.[91] During this conflict, the Navy also played a critical role in supplying U.S. and allied forces and conducting aerial strikes against enemy targets.[92]

In the 1950s, the U.S. Navy began transforming from a steam-powered to a nuclear-powered fleet under the leadership of the controversial and often abrasive Admiral Hyman Rickover (1900–1986). Rickover was particularly influential in developing the U.S. nuclear submarine arsenal, which would go on to incorporate Polaris and Triad missiles as critical components of the U.S. nuclear deterrent, and grappling with a nascent Soviet submarine force.[93]

This decade would also see the Navy challenge the Strategic Air Command's monopoly over strategic bombing, and from 1955 to 1957 the Navy cooperated with the Army on researching a possible liquid-fuel missile capable of being launched from land and surface ships. This research would ultimately produce the Polaris missile first launched off the Florida coast in 1960 by the submarine *U.S.S. George Washington.*[94]

The year 1958 was a particularly busy one for the Navy, and during this time Marines were deployed to Lebanon to protect its government from a possible Syrian invasion, and naval forces were used when the 7th fleet provided aid to Taiwanese forces being shelled by Chinese forces on the islands of Quemoy and Matsu. The most operationally significant event of this year was the Defense Reorganization Act that removed the Chief of Naval Operations (CNO) from operational control of navy fleets, although this legislation allowed the CNO to control planning operations and set operational parameters for naval operations.[95]

The early 1960s saw the Navy play a key role during the 1962 Cuban Missile Crisis, when it enforced a quarantine against Soviet ships that were attempting to resupply the nuclear missiles installed in Cuba.[96] During the remainder of this decade, the Navy was involved in the Vietnam War by providing logistical support for U.S. forces, engaging in air strikes against North Vietnamese and other enemy targets, and conducting riverine and littoral operations against opposing forces.[97]

In the post-Vietnam period, the attention of U.S. Navy doctrine planners was drawn toward the Soviet Navy's growing power and reach. This force, which had been traditionally limited to waters contiguous to Soviet territory, now began to expanding its reach and power projection to multiple global areas. Under the assertive leadership of Admiral Sergei Gorshkov (1910–1988), the Soviet Navy began developing a conventional blue water fleet with ports of call in territories of Soviet allies as diverse as Angola, Cuba, South Yemen, and Vietnam. The Soviets also developed a nuclear submarine fleet that tracked U.S. submarine movements on a global basis and included significant nuclear missile capability that could be deployed against U.S. or allied targets on short notice.[98]

This growing Soviet fleet caused U.S. naval theorists to begin rethinking their views that effective deterrence required the confrontation of potential adversaries with explicit threats of escalation to nuclear war. These theorists now began believing that conventional naval warfare would occur as frequently in the future as in the past, and that it would likely be of greater range and complexity than before. The increasing likelihood of conventional force was due to greater escalatory flexibility as opposed to nuclear force escalation, and it reemphasized the importance of extended conventional war due to the critical importance that economic and industrial strength would play in such conflicts.[99]

Between 1970 and 1974, Chief of Naval Operations (CNO) Admiral Elmo Zumwalt (1920–2000) established a Navy Net Assessment Group to measure U.S. naval effectiveness compared to that of the Soviet Navy. Zumwalt also worked with Admiral Stansfield Turner, who served as Naval War College President from 1972 to 1974, to revise that institution's curriculum to strengthen naval officers' strategic planning abilities. *Future Maritime Strategy Study* (1973), which was released by the Naval War College and the CNO, provided further discussion and analysis of then-current naval strategic trends.[100]

Additional debate within the Navy and DOD's upper echelons also focused on the appropriate size of the U.S. Navy to counter the growing Soviet fleet. This debate produced a wide range of estimates, with a goal of 575 ships set by Secretary of Defense James Schlesinger in 1975; a goal of 600 by his successor Donald Rumsfeld in 1976; and 425–500 by Harold Brown in 1977–1978. The latter figure reflected the Carter Administration's policy that the Navy's surface fleet be designed for peacekeeping operations and for conflicts the Soviet Union chose not to participate in, while still seeking to maintain qualitative U.S. naval superiority.[101]

Concern over the growth of the Soviet Navy and a desire to educate the public and lobby Congress for additional naval funding led the Navy CNO to issue a series of reports called *Understanding Soviet Naval Developments* between 1974 and 1991; these reports provided exhaustive and illustrated analysis of Soviet naval force structure, development, and doctrine for conventional and nuclear forces.[102] These efforts and concern over the status of the U.S. military in relationship to the Soviet Union would pay off with Ronald Reagan's 1980 election and U.S. defense spending increases that would directly benefit the Navy in subsequent years.[103]

The 1980s saw the Reagan Administration attempt to develop a more assertive doctrinal strategy to augment an expanded Navy whose goal was 600 ships. One program developed by the Pacific Fleet during the late 1970s and early 1980s was Project Sea Strike. This program sought to place the Pacific Fleet within a global U.S. naval strategy that would be used if war occurred with the Soviet Union. Sea Strike sought to augment the existing defense-only war plans for this region with offensive capabilities. One Sea Strike provision called for offensive strikes against Soviet bases in the Kamchatka Peninsula and eastern Siberia, and considered offensive operations in the Indian Ocean and Southwest Asia.[104]

Sea Strike also called for taking offensive action against Petropavlovsk, Valdivostok, and the Kuriles with four aircraft carriers that would conduct two waves of air strikes with 100 strike aircraft over the target. Proponents of this operation believed that such strikes would degrade the Soviets' ability to transport forces to Europe to fight against U.S. and NATO forces, enable Chinese forces to be deployed in ways that would restrict Soviet mobility, protect Alaska and the West Coast, and influence Japan to permit U.S. forces to use Japanese bases for additional strikes on Soviet Asia.[105]

The most vivid demonstration of the Reagan Administration's more assertive naval doctrine was the issuance of its 1986 Maritime Strategy. This strategy emphasized offensive fleet engagement preeminence and argued that a nuclear war could be avoided by fighting a protracted global conventional war in which sea control and attrition would be advantageous to the United States and its allies.[106]

It sought to make a naval victory over the Soviets attainable by destroying as many Soviet submarine-launched ballistic missiles as possible, consequently reducing the strategic nuclear threat to the United States, launching strikes against Soviet targets from U.S. carriers, and confining the Soviet fleet to static defensive operations in northern waters. Such U.S. military actions would minimize threats to reinforce efforts to resupply Western Europe by sea. Maritime Strategy critics contended that it could escalate crises by possibly tempting Soviet leaders to use their submarine missiles earlier than intended for fear of losing them to U.S. submarine attack. Strategy proponents countered by saying that allied naval forces had a diverse range of capabilities, such as maintaining presence, conducting surveillance, delivering air or naval strikes, and being deployed or withdrawn depending on existing and evolving strategic situations.[107]

The Cold War's end and the Soviet Union's collapse saw the Navy's bid for global strategic leadership dissipate, although it played an important role in Operation Desert Storm. *From the Sea,* a 1992 Navy White Paper, emphasized the transition from open-ocean war-fighting to joint operations with other armed service branches originating in the sea as well as littoral warfare and maneuver. *From the Sea* also emphasized the criticality of sealift in providing the infrastructure to deliver joint forces and enable them to fight effectively in a major crisis, and argued for the need to flexibly tailor U.S. forces to meet national needs, achieve air, land, and sea battle-space dominance, and establish a Naval Doctrine Command to integrate training and doctrine for regional and littoral war-fighting environments.[108]

The middle 1990s and beyond also saw publication of the Navy's current corpus of keystone doctrinal publications, Navy Doctrinal Publications (NDP), which include *NDP 1 Naval Warfare* (1994), *NDP 2 Naval Intelligence* (n.d.), *NDP 4 Naval Logistics* (2001), *NDP 5 Naval Planning* (n.d.), and *NDP 6 Naval Command and Control* (1995).[109]

March 1997 saw the publication of an updated edition of *From the Sea* entitled *Forward...From the Sea: The Navy Operational Concept.* This document stressed that the *raison d'être* of forward-deployed U.S. naval forces was to project power from the sea to influence events in the world's littoral regions in peace, crisis, and war. Emphasizing that 75 percent of the earth's population and a comparable portion of its major commercial centers are in littoral regions, this document stressed the Navy's peacetime engagement activities, deterrence and conflict prevention objectives, and its determination to fight and win naval conflicts if required. It also stressed the Marines Operational Maneuver from the Sea (OMFTS), which uses highly integrated air, land, and sea operations to carry out amphibious expeditionary operational objectives.[110]

The Navy has played a less significant role in post–9/11 U.S. military operations in Afghanistan and Iraq than has the Army or Marines. However, it is still working to find the right balance of doctrinal thinking to cope with its multiple responsibilities in areas such as maritime security, homeland security, littoral operations, surface warfare, nuclear submarines, and aerial power projection. It is conducting these activities while facing acute fiscal limitations and at a time when more military spending is being devoted to protecting U.S. forces that are engaged in existing combat theaters.[111]

Since U.S. national and economic security depends on secure global oceans, the United States has placed significant emphasis on upgrading its maritime security capabilities. December 2004 saw President George W. Bush direct the Secretaries of Defense and Homeland Security to develop a comprehensive *National Strategy for Maritime Security,* which was issued in September 2005. This document emphasized that threats to national maritime security come from other nations, terrorists, transnational crime, piracy, environmental destruction, and illegal seaborne immigration such as human smuggling. It went on to state that key U.S. strategic maritime security objectives included preventing terrorism and other hostile acts, protecting maritime-related population centers and critical infrastructures such as ports, minimizing damage and expediting recovery, and safeguarding the oceans and their resources.[112]

National Strategy for Maritime Security also commits the United States to increasing international cooperation against maritime threats through intelligence and law enforcement information sharing, expanding the United States' ability to prescreen international cargo before lading, offering maritime and port security training and consultation, embedding security into commercial practices to reduce vulnerabilities and enhance commerce, and deploying layered security to unify public and private security measures against transnational threats.[113]

The most recent U.S. naval strategic document is *A Cooperative Strategy for 21st Century Seapower.* Issued in October 2007, this document bears the imprimatur of the Navy, Marine Corps, and Coast Guard and strives to integrate seapower with other national power elements, emphasizing cooperation with allied nations. *A Cooperative Strategy* stresses that the world's oceanic and littoral regions support 90 percent of world trade and that the United States seeks to "apply seapower in a manner that protects U.S. vital interests even as it promotes greater collective security, stability, and trust."[114]

It stresses that while globalization has increased prosperity in many nations, it has also caused increased resource and capital competition between economic powers, transnational corporations, and international organizations. This has increased popular expectations and may encourage nations to claim expanded sovereignty over oceans, waterways, and natural resources, which may produce conflict. Globalization has also increased information and weapons technology proliferation and enhanced the ability of nations and transnational organizations to challenge maritime access, escape accountability for attacks, and manipulate public perception. Additionally, asymmetric technology use poses threats to the United States and its allies and may involve nuclear and other mass destruction weapons and ferociously destructive attacks on computer, financial, and legal systems. Social instability and climate change may also increase conflict possibilities through storms, arable land loss, and coastal flooding.[115]

The U.S. Navy will respond to these threats by taking the following steps to advance its security interests and those of its allies in achieving heightened global maritime stability:

- Limit regional conflict with forward deployed and decisive maritime power;
- Deter wars between major powers and win national wars;
- Contribute to homeland defense in depth;
- Foster and sustain cooperative relationships with international partners;
- Prevent or contain local disruptions before they have global impact;
- Enhance awareness of maritime domain threats through expanded intelligence, surveillance, and reconnaissance capabilities.[116]

Emerging strategic and doctrinal issues that Navy policymakers must confront include the handling of sea lines of communication (SLOC) security, chokepoints, and their vulnerability to piracy and terrorism in areas such as the Horn of Africa and Straits of Malacca;[117] the implications of ice-free Arctic seas due to climate change and competition for oil and other natural resources involving Russia, the United States, and other countries;[118] how future weapons systems and technologies may affect the Navy's ability to fulfill operational mission mandates;[119] whether China will remain a localized coastal East Asian maritime force or whether it will seek to build a blue water navy capable of challenging U.S. naval preeminence in the Western Pacific and elsewhere;[120] and many other issues covering conventional, nuclear, and other naval and maritime force operational aspects such as sea-based missile defense.

Conclusion

The U.S. military has developed a significant corpus of doctrinal literature to analyze, explain, and rationalize why it has historically conducted military operations, how such operations are currently conducted, and how it plans to conduct military operations in the future.

Military personnel, civilian scholars, and policy analysts provide diverse assessments of the quality of U.S. military doctrine. Much of this analysis and the doctrinal documents themselves are publicly accessible on the Internet or through the substantive historical collections held by many academic research libraries.

Debate over the future directions of U.S. military doctrine and national security strategy will continue as the United States and international military doctrine communities analyze ongoing and future operations in Afghanistan and Iraq.[121]

Such debate will also cover potential future operations that may involve conflict in space, combat against terrorist groups and transnational maritime pirates, information warfare, and potential international crisis situations involving conventional, nonconventional, or weapons of mass destruction operations against countries as diverse as China, Iran, North Korea, and Venezuela.

One arena of military doctrine that is already being debated in scholarly literature and military-oriented blogs by individuals such as Gian Gentile, David Kilcullen, John Nagl, and David Petraeus is whether U.S. military doctrine and war-fighting preparation should focus exclusively on preparing for counterinsurgency operations like those used in Afghanistan and Iraq, continue to emphasize preparations for conventional, nuclear, and other weapons of mass destruction, or seek to combine both of these visions of war-fighting with appropriate doctrine and rules of engagement.[122]

Discussion and analysis of historical, current, and emerging U.S., foreign, and international military doctrine documents and trends is vitally important for those who wish to understand the connection between military action and policymaking, national security and international security policymaking, and why the United States and other countries conduct military operations as they do given the political, diplomatic, economic, legal, normative, and military constraints in which they operate.

The author hopes this book will facilitate greater study and understanding of military doctrine and its accompanying documentation and literature and the importance of this literature in studying military history, political science, international politics, and emerging national security policymaking issues.

Notes

1. See United States Department of State, *Foreign Relations of the United States, 1950: National Security Affairs, Foreign Economic Policy Volume 1* (Washington, DC: Government Printing Office, 1977), 126–492 for *NSC-68* and its documentary trail. *NSC-68* reviews include S. Nelson Drew, ed., *NSC 68: Forging the Strategy of Containment/With Analyses by Paul H. Nitze* (Washington, DC: National Defense University, 1994); David Fautua, "The 'Long-Pull' Army, NSC-68, the Korean War, and the Creation of the Cold War U.S. Army," *Journal of Military History* 61, no. 1 (1997): 93–120; Stephen Casey, "Selling NSC-68: The Truman Administration, Public Opinion, and Politics of Mobilization, 1950–51," *Diplomatic History* 29, no. 4 (2005): 655–690; and John Lewis Gaddis, *Strategies of Containment: A Critical Appraisal of American National Security Policy During the Cold War* (New York: Oxford University Press, 2005).

2. See Jerald A. Combs, "The Compromise That Never Was: George Kennan, Paul Nitze, and the Issue of Conventional Deterrence in Europe, 1949–1952," *Diplomatic History* 15 (1991): 361–386; Christoph Bluth, "Reconciling the Irreconcilable: Alliance Politics and the Paradox of Extended Deterrence in the 1960s," *Cold War History* 1, no. 2 (2001): 73–102; and Andrew M. Johnston, *Hegemony and Culture in the Origins of NATO First-Use, 1945–1955* (New York: Palgrave Macmillan, 2005).

3. For analysis of SIOP, see Peter Pringle and William Arkin, *SIOP: The Secret U.S. Plan for Nuclear War* (New York: Norton, 1983); David Alan Rosenberg, "The Origins of Nuclear

Overkill: Nuclear Weapons and American Strategy, 1945–1960," *International Security* 7, no. 4 (1983): 3–71; William Burr, ed., *The Creation of SIOP-62: More Evidence on the Origins of Overkill* (Washington, DC: National Security Archive, 2004), http://www.gwu.edu/~nsarchiv/NSAEBB/NSAEBB130/ (accessed October 31, 2008); and John H. Rubel, *Doomsday Delayed: USAF Strategic Weapons Doctrine and SIOP-62, 1959–1962. Two Cautionary Tales* (Lanham, MD: Hamilton Books, 2008). For the timetable of SIOP updates until 1990, see Desmond Ball and Robert C. Toth, "Revising the SIOP: Taking War-Fighting to Dangerous Extremes," *International Security* 14, no. 4 (1990): 67.

4. For assessments of massive retaliation, see Great Britain, Ministry of Defense, Chiefs of Staff Committee, Joint Planning Staff, *The Most Effective Pattern of NATO Military Strength for the Next Few Years* (London: Ministry of Defense, 1954); Henry Kissinger, *Nuclear Weapons and Foreign Policy* (New York: Council of Foreign Relations, 1957); Samuel F. Wells Jr., "The Origins of Massive Retaliation," *Political Science Quarterly* 96, no. 1 (1981): 31–52; and H. W. Brands Jr., "Testing Massive Retaliation: Credibility and Crisis Management in the Taiwan Strait," *International Security* 12, no. 4 (1988): 124–151.

5. Ivo H. Daalder, *The Nature and Practice of Flexible Response: NATO Strategy and Theater Nuclear Forces Since 1967* (New York: Columbia University Press, 1991); Beatrice Heuser, *NATO, Britain, France, and the FRG: Nuclear Strategies and Forces for Europe, 1949–2000* (New York: St. Martin's Press, 1997); and Francis J. Gavin, "The Myth of Flexible Response: United States Strategy in Europe During the 1960s," *International History Review* 23 (1975): 847–875.

6. Terry Terriff, *The Nixon Administration and the Making of U.S. Nuclear Strategy* (Ithaca, NY: Cornell University Press, 1995), 1. Additional Schlesinger Doctrine analyses include Colin S. Gray, "Nuclear Strategy: The Debate Moves On," *Journal of the Royal United Services Institute* 121, no. 1 (1976): 44–50; Stephen J. Cimbala, "War-Fighting Deterrence and Alliance Cohesiveness," *Air University Review* 35, no. 6 (1984): 69–73; and William Burr, "The Nixon Administration, the 'Horror Strategy,' and the Search for Limited Nuclear Options, 1969–1972: Prelude to the Schlesinger Doctrine," *Journal of Cold War Studies* 7, no. 3 (2005): 34–78.

7. See Terriff, *The Nixon Administration,* 1–17, and U.S. National Security Council, *National Security Decision Memorandum 242: Policy for Planning the Employment of Nuclear Weapons* (Washington, DC: Nixon Presidential Library and Museum, 1974), 1–5, http://nixon.archives.gov/virtuallibrary/documents/nsdm/nsdm_242.pdf (accessed November 3, 2008).

8. See U.S. National Security Council, *Presidential Directive 59: Nuclear Weapons Employment Policy* (Atlanta: Jimmy Carter Presidential Library, 1980), 2–3, http://www.jimmycarterlibrary.org/documents/pddirectives/pd59.pdf (accessed November 3, 2008); Milton Leitenberg, "Presidential Directive (PD.) 59: United States Nuclear Weapon Targeting Policy," *Journal of Peace Research* 18, no. 4 (1981): 309–317; and Jeffrey Richelson, "PD-59, NSDD-13, and the Reagan Strategic Modernization Program," *Journal of Strategic Studies* 6, no. 2 (1983): 125–146.

9. See Donald Baucom, *The Origins of SDI: 1944–1983* (Lawrence: University Press of Kansas, 1992); and Bert Chapman, *Space Warfare and Defense: A Historical Encyclopedia and Research Guide* (Santa Barbara, CA: ABC-CLIO, 2008), 116–125.

10. Paul Lettow, "President Reagan's Legacy and U.S. Nuclear Weapons Policy," (Washington, DC: The Heritage Foundation, 2006), 4, http://www.heritage.org/Research/NationalSecurity/upload/hl_953.pdf (accessed November 4, 2008).

11. There is extensive literature, representing diverse perspectives, on Reagan Administration nuclear doctrine, national security policymaking, and the end of the Cold War.

Reagan Administration National Security Council directives, some of which remain classified, can be found at Federation of American Scientists, "Presidential Directives and Executive Orders," http://www.fas.org/irp/offdocs/direct.html (accessed November 4, 2008). For another documentary anthology, see William Burr and Robert Wampler, *The Master of the Game: Paul H. Nitze and U.S. Cold War Strategy from Truman to Reagan* (Washington, DC: National Security Archive, 2004), http://www.gwu.edu/~nsarchiv/NSAEBB/NSAEBB139 (accessed November 4, 2008). See also Daniel Wirls, *Buildup: The Politics of Defense in the Reagan Era* (Ithaca, NY: Cornell University Press, 1992); Peter W. Rodman, *More Precious Than Peace: The Cold War and the Struggle for the Third World* (New York: C. Scribner's Sons, 1994); Mark P. Lagon, *The Reagan Doctrine: Sources of American Conduct in the Cold War's Last Chapter* (Westport, CT: Praeger, 1994); Ralph Summy and Michael E. Salla, eds., *Why the Cold War Ended: A Range of Interpretations* (Westport, CT: Greenwood Press, 1995); and James M. Scott, *Deciding to Intervene: The Reagan Doctrine and American Foreign Policy* (Durham, NC: Duke University Press, 1996).

12. See U.S. National Security Council, *National Security Directive 70: United States Nonproliferation Policy* (College Station, TX: George Bush Presidential Library, 1992), 3–4, http://bushlibrary.tamu.edu/research/pdfs/nsd/nsd70.pdf (accessed November 4, 2008). See also William W. Newman, "The Structures of National Security Decision Making: Leadership, Institutions, and Politics, in the Carter, Reagan, and G.H.W. Bush Years," *Presidential Studies Quarterly* 34 (2004): 272–306.

13. U.S. Department of Energy, Nevada Operations Office, *United States Nuclear Tests July 1945 Through September 1992* (Las Vegas: DOE Nevada Operations Office, 2000), vii. For literature on the unsuccessful Comprehensive Test Ban Treaty ratification, see U.S. Congress, Senate Committee on Armed Services, *Comprehensive Test Ban Treaty* (Washington, DC: Government Printing Office, 2000) and U.S. Congress, Senate Committee on Foreign Relations, *Final Review of the Comprehensive Nuclear Test Ban Treaty (Treaty Doc. 105–28)* (Washington, DC: Government Printing Office, 2000).

14. See Peter D. Zimmerman and David W. Dorn, *Computer Simulation and the Comprehensive Test Ban Treaty* (Washington, DC: Center for Technology and National Security Policy, National Defense University, 2002); U.S. Department of Energy, National Nuclear Security Administration, Nevada Site Office, *Stockpile Stewardship Program* (Las Vegas: Nevada Site Office, 2004), http://www.nv.doe.gov/library/FactSheets/DOENV_1017.pdf (accessed November 4, 2008); and Gene Aloise, *Nuclear Weapons: Preliminary Results of Review of Campaigns to Provide Scientific Support for the Stockpile Stewardship Program* (Washington, DC: U.S. Government Accountability Office, 2005).

15. Nuclear Threat Initiative, "U.S. Nuclear Posture Reviews," (n.d.), http://www.nti.org/f_wmd411/f2c/html (accessed November 4, 2008).

16. U.S. Department of Defense, *Nuclear Posture Review Report* (Washington, DC: Department of Defense, 2002), 1–3, http://www.defenselink.mil/news/Jan2002/d20020109npr.pdf (accessed November 4, 2008).

17. U.S. Department of Energy and Department of Defense, *National Security and Nuclear Weapons in the 21st Century* (Washington, DC: Department of Energy and Department of Defense, 2008), 7–8.

18. Ibid., 11–16.

19. Ibid., 18–22. See also Jonathan Medalia, *The Reliable Replacement Warhead Program: Background and Current Developments* (Washington, DC: Library of Congress, Congressional Research Service, 2008).

20. See James A. Mowbray, "Air Force Doctrine Problems 1926–Present," *Airpower Journal* 9, no. 4 (1995): 22 and Carl H. Builder, *The Icarus Syndrome: The Role of Air Power Theory in the Evolution of the U.S. Air Force* (New Brunswick, NJ: Transaction, 2003), 76–79.

21. Mowbray, "Air Force Doctrine," 27; and Robert Frank Futrell, *Ideas, Concepts, Doctrine: Basic Thinking in the United States Air Force 1907–1960: Volume I* (Maxwell Air Force Base, AL: Air University Press, 1989), 145–171.

22. Mowbray, "Air Force Doctrine," 28, and Futrell, *Ideas: Vol. I,* 206–208.

23. Futrell, *Ideas: Vol. I,* 393.

24. Mowbray, "Air Force Doctrine," 29.

25. See Builder, *The Icarus Syndrome,* 165–177; and F. W. Jennings, "Doctrinal Conflict Over the Word Aerospace," *Airpower Journal* 4, no. 3 (1990): 46–59.

26. Mowbray, "Air Force Doctrine," 31–32; and Futrell, *Ideas, Concepts, Doctrine: Basic Thinking in the United States Air Force 1961–1984: Volume II* (Maxwell Air Force Base, AL: Air University Press, 1989), 230–235.

27. Mowbray, "Air Force Doctrine," 32–3. Also, see Earl H. Tilford, *Setup: What the Air Force Did in Vietnam and Why* (Maxwell Air Force Base, AL: Air University Press, 1991) for a review of Air Force doctrine and strategy during Vietnam.

28. Benjamin S. Lambeth, *The Transformation of American Air Power* (Ithaca, NY: Cornell University Press, 2000), 48–49.

29. Futrell, *Ideas: Vol. II,* 744.

30. Lambeth, *Transformation,* 103–152.

31. United States Department of the Air Force, *Gulf War Air Power Survey,* 5 vols. (Washington, DC: Dept. of the Air Force, 1993), http://www.airforcehistory.hq.af.mil/Publications/Annotations/gwaps.htm (accessed November 6, 2008).

32. Johnny R. Jones, *Development of Air Force Basic Doctrine, 1947–1992* (Maxwell Air Force Base, AL: Air University Press, 1997), 31–32, 61–63.

33. Chapman, *Space Warfare and Defense,* 41–43.

34. Air University, Center for Strategy and Technology, *Welcome to Air Force 2025,* http://csat.au.af.mil/2025/ (accessed November 6, 2008).

35. Lambeth, *Transformation,* 223–226.

36. U.S. Air Force, *Air Force Basic Doctrine AFDD 1* (Washington, DC: U.S. Air Force, 2003), iii–iv.

37. A current list of Air Force doctrinal documents can be found at Air Force Publishing, http://www.e-publishing.af.mil/.

38. For introductions to this proliferating literature, see Robert A. Doughty, *The Evolution of U.S. Army Tactical Doctrine, 1946–1976* (Fort Leavenworth, KS: U.S. Army Command and General Staff College, 1979); Andrew James Birtle, *U.S. Army Counterinsurgency and Contingency Operations Doctrine, 1942–1976* (Washington, DC: Center of Military History, U.S. Army, 2006); Brian McAllister Linn, *The Echo of Battle: The Army's Way of War* (Cambridge, MA: Harvard University Press, 2007); and Ingo Trauschweizer, *The Cold War U.S. Army: Building Deterrence for Limited War* (Lawrence: University Press of Kansas, 2008).

39. See Trauschweizer, *Cold War,* 81–113 and Andrew J. Bacevich, *Pentomic Era: The U.S. Army between Korea and Vietnam* (Washington, DC: National Defense University Press, 1986).

40. See Robert A. Doughty, *Evolution,* 22; Trauschweizer, *Cold War,* 114–161; and Trauschweizer, "Learning with an Ally: The U.S. Army and the Bundeswehr in the Cold War," *Journal of Military History* 72 (2008): 489–490.

41. Trauschweizer, *Cold War,* 180–185.

42. Examinations of the Army's Vietnam failures include Doughty, *Evolution,* 29–40; Harry G. Summers Jr., *On Strategy: A Critical Analysis of the Vietnam War* (New York: Dell Books, 1982); Bruce Palmer Jr., *The 25-Year War: America's Military Role in Vietnam* (New York: Touchstone Books, 1984); Julian J. Ewell and Ira A. Hunt Jr., *Sharpening the Combat Edge: The Use of Analysis to Reinforce Military Judgement* (Washington, DC: Department of the Army, 1995); and John A. Nagl, *Learning to Eat Soup with a Knife: Counterinsurgency Lessons from Malaya and Vietnam* (Chicago: University of Chicago Press, 2005), 115–223.

43. For TRADOC's official history, see Anne Chapman et al., *Transforming the Army: TRADOC's First Thirty Years, 1973–2003* (Fort Belvoir, VA: Defense Technical Information Center, 2003).

44. See Henry G. Cole, *General William E. DePuy: Preparing the Army for Modern War* (Lexington: University Press of Kentucky, 2008) and Richard Lock-Pullan, "'An Inward Looking Time, 1973–1976': The United States Army, 1973–1976," *Journal of Military History* 67 (2003): 483–512.

45. George Q. Flynn, *The Draft, 1940–1973* (Lawrence: University Press of Kansas, 1993).

46. See Lock-Pullan, "Inward Looking Time," 497 and Trauschweizer, "Learning with an Ally," 496. For more on the perspective that the Army sought to avoid or ignore Vietnam's lessons on counterinsurgency warfare's importance, see Conrad C. Crane, *Avoiding Vietnam: The U.S. Army's Response to Defeat in Southeast Asia* (Carlisle Barracks, PA: Strategic Studies Institute, U.S. Army War College, 2002).

47. Trauschweizer, "Learning with an Ally," 497.

48. Lock-Pullan, "Inward Looking Time," 498–499.

49. See Doughty, *Evolution,* 41–42 and Headquarters, Department of the Army, *FM 100–5 Operations* (Washington, DC: Headquarters, Department of the Army, 1976), 1–1.

50. See Doughty, *Evolution,* 43 and John L. Romjue, "The Evolution of the Airland Battle Concept," *Air University Review* 35, no. 4 (1984): 4.

51. Lock-Pullan, "Inward Looking Time," 507–508.

52. Trauschweizer, *Cold War,* 215.

53. See Trauschweizer, "Learning with an Ally," 501–502 and Romjue, "Evolution," 9.

54. Trauschweizer, *Cold War,* 222. For the text of the updated *FM 100–5* that incorporates AirLand Battle, see Headquarters, Department of the Army, *FM 100–5 Operations* (Washington, DC: Headquarters, Department of the Army, 1982).

55. Romjue, "Evolution," 10, 12.

56. See Trauschweizer, *Cold War,* 228 and Stephen A. Bourque, *Jayhawk!: The VII Corps in the Persian Gulf War* (Washington, DC: Department of the Army, 2002), 455–461.

57. U.S. Congress, Office of Technology Assessment, "*New Technology for NATO: Implementing Follow-On Forces Attack* (Washington, DC: Government Printing Office, 1987), 3–4.

58. U.S. Army, *FM 100–5 Operations* (Washington, DC: Headquarters, Department of the Army, 1993), 1–1 to 1–5.

59. Literature on Army peacekeeping operations during the 1990s and debate over the desirability or feasibility of Army peacekeeping doctrine includes Jennifer Morrison Taw and John E. Peters, *Operations Other Than War: Implications for the U.S. Army* (Santa Monica, CA: Rand Corporation, 1995); Max G. Manwaring, "Peace and Stability: Lessons from Bosnia," *Parameters* 28, no. 4 (1998/1999): 28–38; Mark Bowden, *Black Hawk Down: A Story of Modern War* (New York: Atlantic Monthly Press, 1999); John Davis and Howard Olsen, *Training U.S. Army Officers for Peace Operations: Lessons From Bosnia* (Washington, DC: U.S. Institute of Peace, 1999); Norman L. Cooling, "Operation Restore Hope in

Somalia: A Tactical Action Turned Strategic Defeat," *Marine Corps Gazette* 85, no. 9 (2001): 92–106; and Robert M. Cassidy, *Peacekeeping in the Abyss: British and American Peacekeeping Doctrine and Practice after the Cold War* (Westport, CT: Greenwood Publishing, 2004).

60. See Steven Metz, *Counterinsurgency: Strategy and the Phoenix of American Capability* (Carlisle Barracks, PA: U.S. Army War College, Strategic Studies Institute, 1995); John A. Nagl, "Counterinsurgency in Vietnam: American Organizational Culture and Learning," in *Counterinsurgency in Modern Warfare,* eds. Daniel Marston and Carter Malkasian (Westminster, MD: Osprey Publishing, 2008), 146–148; and Trauschweizer, "Learning with an Ally," 507–508.

61. Examples of works examining U.S. Army attempts to develop appropriate doctrine for counterinsurgency operations in these conflicts include Vince Crawley, "High-Speed Warfare: Combat in Iraq is Driving New Doctrines and Propelling Transformation," *Air Force Times* 64, no. 27 (2004): 18; Christopher Hickey, "Principles and Priorities in Training for Iraq," *Military Review* 87, no. 2 (2007): 27–32; Nathan Hodge, "U.S. Draws on Experience in Afghanistan and Iraq to Shape Counterinsurgency Manual," *Jane's International Defence Review* 40, no. 10 (2007): 10; Gian P. Gentile, "The Dogmas of War: A Rigid Counterinsurgency Doctrine Obscures Iraq's Realities," *Armed Forces Journal* 145, no. 5 (2007): 38–40; Joseph R. Cerami and Jay W. Biggs, eds., *The Interagency and Counterinsurgency Warfare: Stability, Security, Transition, and Reconstruction Roles* (Carlisle Barracks, PA: Strategic Studies Institute, U.S. Army War College, 2007); David M. Tressler, *Negotiation in the New Strategic Environment: Lessons from Iraq* (Carlisle Barracks, PA: Strategic Studies Institute, U.S. Army War College, 2007); and Peter R. Mansoor, *Baghdad at Sunrise: A Brigade Commander's War in Iraq* (New Haven, CT: Yale University Press, 2008).

62. U.S. Army, *The Army: Our Army at War: Relevant and Ready Today and Tomorrow* (Washington, DC: Headquarters, Department of the Army, 2005), 2–3.

63. Ibid., 2–2.

64. See U.S. Army, *FM 3–24 Counterinsurgency* (Washington, DC: Headquarters, Department of the Army, 2006), i–v. For Petraeus's role in writing *FM–24,* see John Nagl, "The Evolution and Importance of Army/Marine Corps Field Manual 3–24, Counterinsurgency," Small Wars Journal Blog, June 27, 2007, http://smallwarsjournal.com/blog/2007/6/the-evolution-and-importance-o/ (accessed November 11, 2008); Frank Hofman, "Neo-Classical Counterinsurgency?," *Parameters* 37, no. 2 (2007): 71–87; and Sheila Miyoshi Jager, *On the Uses of Cultural Knowledge* (Carlisle Barracks, PA: Strategic Studies Institute, U.S. Army War College, 2007).

65. See U.S. Marines Corps, *Small Wars Manual* (Washington, DC: Government Printing Office, 1940), http://www.smallwars.quantico.usmc.mil/SWM/1215.pdf (accessed November 12, 2008); Ronald Schaffer, "The 1940 Small Wars Manual and the 'Lessons of History'," *Military Affairs* 36, no. 2 (1972): 46–51; David Keithly and Paul Melshin, "Past as Prologue: USMC Small Wars Doctrine," *Small Wars and Insurgencies* 8, no. 2 (1997): 87–108; and David J. Ulbrich, "Revisiting Small Wars: A 1933 Questionnaire, Vernon E. Megee, and the Small Wars Manual," *Marine Corps Gazette* 90, no. 11 (2006): 74–75.

66. See Victor Krulak, *First to Fight: An Inside View of the U.S. Marine Corps* (Annapolis, MD: Naval Institute Press, 1984) and Terry Terriff, "'Innovate or Die': Organizational Culture and the Origins of Maneuver Warfare in the United States Marine Corps," *The Journal of Strategic Studies* 29, no. 3 (2006): 480–484.

67. Kenneth J. Clifford, *Progress and Purpose: A Developmental History of the United States Marine Corps, 1900–1970* (Washington, DC: History and Museums Division Headquarters, United States Marine Corps, 1973), 71–72.

68. Public Law 82–416, 66, *U.S. Statutes at Large*, 283.

69. See Clifford, *Progress and Purpose*, 83–85 and Charles R. Smith, ed., *The U.S. Marines in the Korean War* (Washington, DC: History Division, U.S. Marine Corps, 2007).

70. Allan R. Millett, *Semper Fidelis: The History of the United States Marine Corps* (New York: Simon and Schuster, 1991), 545.

71. See Clifford, *Progress and Purpose*, 97–113 and Keithly and Melshin, "Past as Prologue," 100; also, for one of the many official Marine Corps Vietnam War histories, see Jack Shulimson, *U.S. Marines in Vietnam: An Expanding War, 1966* (Washington, DC: History and Museums Division, U.S. Marine Corps, 1982).

72. See Michael A. Hennessy, *Strategy in Vietnam: The Marines and Revolutionary Warfare in I Corps, 1965–1972* (Westport, CT: Praeger, 1997), 181 and Terriff, "Innovate or Die," 485.

73. Ibid., 485–489.

74. Martin Binkin and Jeffrey Record, *Where Does the Marine Corps Go from Here?* (Washington, DC: Brookings Institution, 1976), 71–86.

75. Terriff, "Innovate or Die," 489.

76. Millett, *Semper Fidelis*, 547.

77. United States, DOD Commission on Beirut International Airport Terrorist Act, October 23, 1983, *Report of the DOD Commission on Beirut International Airport Terrorist Act, October 23, 1983* (Washington, DC: Government Printing Office, 1984).

78. Erik J. Dahl, "Warning of Terror: Explaining the Failure of Intelligence against Terrorism," *The Journal of Strategic Studies* 28, no. 1 (2005): 31–55.

79. See Kenneth F. McKenzie, Jr., "On the Verge of a New Era: The Marine Corps and Maneuver Warfare," *Marine Corps Gazette* 77, no. 7 (1993): 62–67; Terriff, "Innovate or Die," 475; and Fidelian Dameon, "The Road to FMFM1: The United States Marine Corps and Maneuver Warfare Doctrine, 1979–1989" (master's thesis, Kansas State University, 2008).

80. U.S. Marine Corps, *FMFM 1Warfighting* (Washington, DC: Headquarters, United States Marine Corps, 1989), 29.

81. See Dennis P. Mroczkowski, *U.S. Marines in the Persian Gulf, 1990–1991: With the 2nd Marine Division in Desert Shield and Desert Storm* (Washington, DC: History and Museums Division, Headquarters, U.S. Marine Corps, 1993) and U.S. Marine Corps Intelligence Activity, *Challenges to Naval Expeditionary Warfare* (Washington, DC: The Office, 1997).

82. Analyses of Marine Corps operations in the Global War on Terror include Bob Krum, "Why are the Marines in Afghanistan?," *U.S. Naval Institute Proceedings* 128, no. 1 (2002): 112; Scott E. Broberg, "Are We Properly Prepared for Helicopter Operations in Afghanistan?," *Marine Corps Gazette* 86, no. 5 (2002): 70–74; Matt Hilbrun, "Policing the Insurgents: Marines in Iraq Adapt New Technology and Law Enforcement Tactics," *Sea Power* 49, no. 3 (2006): 44; Nicholas E. Reynolds, *U.S. Marines in Iraq, 2003: Basrah, Baghdad, and Beyond* (Washington, DC: History Division, U.S. Marine Corps, 2007); Timothy J. Bailey, "Why Not Afghanistan?: This Mission is Still to Be Accomplished," *Marine Corps Gazette* 91, no. 8 (2007): 14–17; and James S. Corum, "On Airpower, Land Power, and Counterinsurgency: Getting Doctrine Right," *Joint Force Quarterly* 49 (2008): 93–97.

83. See Frank G. Hoffman, "A Marine Corps for a Global Century: Expeditionary Maneuver Brigades," in *Globalization and Maritime Power*, ed. Sam J. Tangredi (Washington, DC: National Defense University Press, 2002), 427–428 and U.S. Marine Corps, *Marine Corps Strategy 21* (Washington, DC: Headquarters, United States Marine Corps, 2000).

84. U.S. Marine Corps, *USMC Concepts & Programs 2008* (Washington, DC: Headquarters, United States Marine Corps, 2008), 2, 6.

85. Stephen Howarth, *To Shining Sea: A History of the United States Navy, 1775–1991* (New York: Random House, 1991), 480.

86. Robert E. Fisher, "The U.S. Navy's Search for a Strategy, 1945–1947," *Naval War College Review* 48, no. 3 (1995): 73–86.

87. James J. Tritten, *Development Issues for Multinational Navy Doctrine* (Norfolk, VA: Naval Doctrine Command, 1996), 3.

88. Arthur A. Adkins, "Doctrine for Naval Planning: The Once and Future Thing," *Naval War College Review* 49, no. 1 (1996): 66.

89. U.S. Congress, House Committee on Armed Services, *The National Defense Program-Unification and Strategy* (Washington, DC: Government Printing Office, 1949), 521.

90. Jeffrey G. Barlow, *Revolt of the Admirals: The Fight for Naval Aviation, 1945–1950* (Washington, DC: Naval Historical Center, 1994), 294.

91. Curtis A. Utz, *Assault from the Sea: The Amphibious Landing at Inchon* (Washington, DC: Naval Historical Center, 1994).

92. See George W. Baer, *The U.S. Navy, 1890–1990: One Hundred Years of Sea Power* (Stanford, CA: Stanford University Press, 1996), 4; James A. Field Jr., *History of United States Naval Operations: Korea* (Washington, DC: Department of the Navy, Naval Historical Center, 2000); and Malcolm Muir, *Sea Power on Call: Fleet Operations, June 1951–July 1953* (Washington, DC: Naval Historical Center, 2005).

93. For assessments of the importance of the U.S. nuclear submarine program, see Howarth, *To Shining Sea,* 494–497; U.S. Congress, House Committee on Armed Services, Subcommittee on Sea Power and Strategic and Critical Materials, *Report on the United States Nuclear-Powered Submarine Attack Program* (Washington, DC: Government Printing Office, 1979); Francis Duncan, *Rickover and the Nuclear Navy: The Discipline of Technology* (Annapolis, MD: Naval Institute Press, 1990); and Graham Spinardi, *From Polaris to Trident: The Development of US Fleet Ballistic Missile Technology* (New York: Cambridge University Press, 1994).

94. Jakub J. Grygiel, "The Dilemmas of US Maritime Supremacy in the Early Cold War," *The Journal of Strategic Studies* 28, no. 2 (2005): 201.

95. See Baer, *U.S. Navy,* 4; and David Alan Rosenberg, *Arleigh Burke: The Last CNO* (Washington, DC: Naval Historical Center, 2005), 15–17, http://www.history.navy.mil/bios/burke_rosen2.htm (accessed November 19, 2008).

96. Howarth, *To Shining Sea,* 507–508.

97. Edward J. Marolda, *By Sea, Air, and Land: An Illustrated History of the U.S. Navy and the War in Southeast Asia* (Washington, DC: Naval Historical Center, 1994).

98. For coverage of the expanding Soviet naval presence, see U.S. Department of the Navy, Office of the Chief of Naval Operations, *Understanding Soviet Naval Developments,* 4th ed. (Washington, DC: Government Printing Office, 1981), 15–29; Bruce W. Watson and Susan M. Watson, eds., *The Soviet Navy: Strengths and Liabilities* (Boulder, CO: Westview Press, 1986); and Robert Warring Herrick, *Soviet Naval Theory and Policy: Gorshkov's Inheritance* (Newport, RI: Naval War College Press, 1988)

99. John B. Hattendorf, *The Evolution of the U.S. Navy's Maritime Strategy, 1977–1986* (Newport, RI: Naval War College Press, 2004), 6–7.

100. Ibid., 8–9.

101. Ibid., 9.

102. U.S. Department of the Navy, Office of the Chief of Naval Operations, *Understanding Soviet Naval Developments,* 6th ed. (Washington, DC: Government Printing Office, 1991).

103. Howarth, *To Shining Sea,* 538.

104. Hattendorf, *Evolution,* 18.

105. Ibid., 19.

106. Baer, *U.S. Navy,* 5.

107. See James D. Watkins, "The Maritime Strategy," *Proceedings: U.S. Naval Institute* 112, no. 1 (1986): 8 and Christopher A. Ford and David A. Rosenberg, "The Naval Intelligence Underpinnings of Reagan's Maritime Strategy," *The Journal of Strategic Studies* 28, no. 2 (2005): 394–395. Maritime Strategy documents including overall strategy justification, amphibious warfare strategy, and the rationale for a 600-ship navy were published in the January 1986 publication of *Proceedings: U.S. Naval Institute.*

108. See Sean C. O'Keefe, Frank B. Kelso II, and Carl E. Mundy Jr., "From the Sea: A New Direction for the Naval Services," *Marine Corps Gazette* 76, no. 11 (1992): 18–22 and Baer, *U.S. Navy,* 6.

109. These and other service doctrine resources can be found at Defense Technical Information Center, Joint Electronic Library, http://www.dtic.mil/doctrine/.

110. U.S. Navy, Chief of Naval Operations, *The United States Navy: Forward... From the Sea: The Navy Operational Concept* (Washington, DC: Chief of Naval Operations, 1997), 1–10, http://www.chinfo.navy.mil/navypalib/policy/fromsea/fseanoc.html (accessed November 20, 2008).

111. See Naval Studies Board, *Naval Analytical Capabilities: Improving Capabilities-Based Planning* (Washington, DC: National Research Council, 2005); U.S. Congress, House Committee on Armed Services, Projection Forces Subcommittee, *U.S. Navy's Future Submarine Force Structure* (Washington, DC: Government Printing Office, 2006); and U.S. Congressional Budget Office, *Options for the Navy's Future Fleet* (Washington, DC: Congressional Budget Office, 2006) for a representative sampling of literature on future naval force structure options.

112. President of the United States, *National Strategy for Maritime Security* (Washington, DC: White House, 2005), ii, 3–12.

113. Ibid., 14–23.

114. U.S. Navy, *A Cooperative Strategy for 21st Century Seapower* (Washington, DC: U.S. Navy, 2007), 2, http://www.navy.mil/maritime/ (accessed November 20, 2008).

115. Ibid., 3–4.

116. Ibid., 6–13. For assessments of this document, see James Kurth, "The New Maritime Strategy: Confronting Peer Competitors, Rogue States, and Transnational Insurgents," *Orbis* 51, no. 4 (2007): 585–600; Andrew S. Erickson, "Assessing the New U.S. Maritime Strategy: A Window into Chinese Thinking," *Naval War College Review* 61, no. 4 (2008): 53; and related articles by Chinese strategic analysts in the Autumn 2008 publication of *Naval War College Review.*

117. Donna J. Nincic, "Sea Lane Security and U.S. Maritime Trade: Chokepoints as Scarce Resources," in *Globalization and Maritime Power,* ed. Sam J. Tangredi (Washington, DC: National Defense University Press, 2002), 143–169.

118. Jessie C. Carman, "Economic and Strategic Implications of Ice-Free Arctic Seas," in *Globalization and Maritime Power,* ed. Sam J. Tangredi (Washington, DC: National Defense University Press, 2002), 171–188.

119. See Henry H. Gaffney, "The Navy before and after September 11," in *Globalization and Maritime Power,* ed. Sam J. Tangredi (Washington, DC: National Defense University Press, 2002), 535–549, and Geoffrey Till, *Naval Transformation, Ground Forces, and the Expeditionary Impulse: The Sea-Basing Debate* (Carlisle Barracks, PA: Strategic Studies Institute, U.S. Army War College, 2006).

120. For a partial sampling of this topic's burgeoning literature, see Lyle Goldstein, ed., *China's Nuclear Force Modernization* (Newport, RI: Naval War College Press, 2005); David Lei, "China's New Multi-faceted Maritime Strategy," *Orbis* 52, no. 1 (2008): 139–157; Ronald O'Rourke, *China's Naval Modernization: Implications for U.S. Navy Capabilities—Background and Issues for Congress* (Washington, DC: Library of Congress, Congressional Research Service, 2008); and Toshi Yoshihara and James Holmes, "China's New Undersea Nuclear Deterrent: Strategy, Doctrine, and Capabilities," *Joint Force Quarterly* 50 (2008): 31–38.

121. Peter R. Mansoor, *Baghdad at Sunrise.*

122. Those who favor preeminent emphasis on counterinsurgency include Nagl, *Learning to Eat Soup;* David H. Petraeus, "Learning Counterinsurgency: Observations from Soldiering in Iraq," *Military Review* 86, no. 1 (2006): 2–12; and David Kilcullen, "Counter-Insurgency Redux," *Survival* 48, no. 4 (2006): 111–130. West Point historian Gian P. Gentile is a leading figure among those concerned that the military's emphasis on counterinsurgency doctrine is weakening its ability to conduct conventional operations. Examples of his writings include "Eating Soup with a Spoon: Missing from the New COIN Manual's Pages is the Imperative to Fight," *Armed Forces Journal* 145 (September 2007): 30–33, 46; "The Dogmas of War: A Rigid Counterinsurgency Doctrine Obscures Iraq's Realities," *Armed Forces Journal* 145 (December 2007): 38–40; "Our COIN Doctrine Removes the Enemy from the Essence of War," *Armed Forces Journal* 145 (January 2008): 39; and "Misreading the Surge Threatens U.S. Army's Conventional Capabilities," *World Politics Review,* March 4, 2008, 1–4, http://www.worldpoliticsreview.com/article.aspx?id=1715 (accessed November 21, 2008). A summative assessment of this debate can be found in T. X. Hammes, "The Art of Petraeus," *The National Interest* 98 (2008): 53–59.

CHAPTER 2

U.S. Government Military Doctrine Resources

The U.S. Government is the world's leading military doctrine information producer. These resources are produced by many armed service branches and this chapter will primarily focus on publicly accessible Internet resources. It will begin with coverage of joint U.S. military doctrine documents. Joint, as used in military terminology, refers to using two or more armed services of the same nation in coordinated action to obtain common objectives. Joint military cooperation and planning has received major emphasis within the U.S. military as a result of the 1986 Goldwater-Nichols Act. This has compelled the U.S. military to place heavy emphasis on collaborative planning between armed services branches and officers serving in joint commands as a way to diminish inter-service rivalries and promote military career advancement.[1]

This chapter will describe how to find national security strategy documents produced by recent presidential administrations and military doctrine documents produced by individual branches of the U.S. military that contain information about the organizations within the U.S. military responsible for producing, revising, and updating military doctrinal literature. The primary emphasis of this chapter will be on finding current U.S. military doctrinal and national security strategy literature since much of it is accessible through the Internet.

Students of military doctrine documents will be able to find this literature in some of the United States' federal depository libraries. These libraries provide Americans with free access to information produced by the U.S. Government and are paid for with our tax dollars. A directory of federal depository libraries can be found at http://catalog.fdlp.gov/fdlpdir/FDLPdir.jsp. Such documents are most likely to be found in major university libraries and will likely be arranged in the U.S. Government Printing Office's Superintendent of Documents (SuDoc) classification system in which documents are arranged alphabetically by the agency producing the document. Of tangible format (print or microfiche) military doctrine publications since the late 1940s, joint doctrine publications produced by the Joint Chiefs of Staff (JCS) can be found in the D 5.12 SuDoc call number range, Army Field Manuals (FM) are in the D 101.20 range, Navy doctrine publications are in the D 207.402 range, Marine Corps doctrine publications are in the D 214.9/ range, and Air Force Doctrine publications are in the D 301.134 call number range. For earlier doctrinal publications from the various armed services, Army FMs can be found in the W 1.33 and W 3.63 call number ranges, and relevant Navy and Marine Corps publications can be found in the N 1.13 and N 9.9/3 and M 209.8 call number ranges. More recent versions of these documents are likely available on the Internet.[2]

National Security Strategy Documents

The most authoritative national security strategy documents are produced by the White House and National Security Council with collaborative input from other military and government agencies.

They represent declarative policy documents issued by presidential administrations, which reflect then-prevailing administration national security policy objectives and priorities.[3] One of the first of these documents was issued by the Reagan Administration in January 1987 as *National Security Strategy of the United States*. This 41-page document sought to provide "a blueprint for freedom, peace, and prosperity," which it saw as being bulwarks of U.S. national security policy. This strategy included commitment to world freedom, peace, and prosperity; strong and close relationships with global alliance partners; active assistance to those struggling for self-determination, freedom, and reasonable living standards; a willingness to be realistic about the Soviet Union and to make public moral distinctions between democracy and totalitarianism; and a commitment to seeking constructive ways of working with Soviet leaders to prevent war and make the world more peaceful.[4]

This document went on to stress fundamental characteristics of U.S. national security strategy, such as a healthy and growing national economy, U.S. regional security policies in the Western Hemisphere, Soviet Union, and Eastern Europe, the importance of maintaining conventional and nuclear deterrent forces, and the need for the physical capabilities to implement these objectives.[5]

These documents have appeared fairly regularly in subsequent years. The Reagan Administration issued another *National Security Strategy* in January 1988. The George H. W. Bush Administration issued versions of this strategic document in March 1990, August 1991, and January 1993 to cover events such as the Persian Gulf War, the fall of the former Soviet Union, and the emergence of peace-keeping as a potential U.S. national security policy concern.

The two principal Clinton Administration versions of these documents were *A National Security Strategy of Engagement and Enlargement*, released in February 1995, and *A National Security Strategy for a Global Age*, released in December 2000. Topics addressed in the 1995 edition included counterterrorism, drug trafficking, combating weapons of mass destruction proliferation, the North American Free

Trade Agreement, and energy security.[6] Emphases of the 2000 document included seeking to shape the international security environment through diplomacy, economic cooperation, arms control and nonproliferation activities, and military presence and engagement, along with promoting open trade, enhancing American competitiveness, and advancing democracy.[7]

The George W. Bush Administration, influenced by the 9/11 terrorist attacks and subsequent wars in Afghanistan and Iraq, has issued two important national security strategy documents. The 2002 *National Security Strategy of the United States* placed particular emphasis on the fighting of terrorism as a critical concern for U.S. national security policy. Conventional thinking aspects of this document, in comparison with other recent presidential national security documents, stressed championing aspirations for human dignity, strengthening alliances to defeat global terrorism, and working to prevent attacks against the U.S. and its allies. It also expressed the need to work with others to defuse regional conflicts, prevent enemies from threatening the United States and its allies with weapons of mass destruction (WMD), expand global economic growth through free markets and free trade, and expand economic and political development by opening societies and building democratic infrastructures.[8]

The most innovative and controversial provision of this document was its declaration of willingness to take preemptive action against hostility to the United States and its interests by:

> identifying and destroying the threat before it reaches our borders. While the United States will constantly strive to enlist the support of the international community, we will not hesitate to act alone, if necessary, to exercise our right of self-defense by acting preemptively against such terrorists, to prevent them from doing harm against our people and our country.[9]

This militarily prudent policy response to the evolving threat of an agile and amorphous transnational enemy has received considerable criticism, which ranges from hysterical denunciation, pragmatic suggestion for modification, and critical support of its validity.[10] Debate on military preemption and other aspects of the George W. Bush Administration's national security policy will continue for decades.

The 2006 edition of *National Security Strategy* reiterated many of the key emphases of the 2002 document, including preventing terrorist network attacks before they occur, denying WMD to rogue states and terrorist allies who would use such weapons without hesitation, denying terrorist groups the support and sanctuary of rogue states, and denying terrorists control of any nation they would use as a base for launching terror.[11]

Another important series of military documents detailing overall national military strategy is the *Quadrennial Defense Review (QDR)*. This review was issued by the U.S. military in 1997, 2001, and 2006, and the next edition is expected during the opening year of the Obama Administration. This legislation was mandated by Public Law 103–62, the Government Performance and Results Act. This congressional statute requires *QDR* to include the following content:

- Assumed or defined U.S. national security interests that inform national defense strategy;
- Threats to assumed or defined U.S. national security interests, including the readiness of U.S. forces, allied cooperation and mission-sharing, warning times of enemy attacks, engagement levels in operations other than war, and withdrawal from such operations;
- The effect on U.S. force structure and readiness for high-intensity combat preparations, as well as the participation, staffing, and sustainment policies that national defense strategy would require to support a conflict engagement lasting over 120 days;
- Anticipated roles and missions of reserve components in such missions;
- Assessment of the appropriate ratio of combat forces to support forces;
- Examination of strategic and tactical airlift, sealift, and ground transportation capabilities to support national defense strategy, including forward presence and pre-deployment capabilities;
- The extent to which resources may need to be shifted to two or more combat theaters in the event of conflict in such theaters; and
- How force structure will be impacted by technologies anticipated to become available in the next 20 years.[12]

The 1997, 2001, and 2006 *QDRs* are accessible at http://www.defenselink.mil/qdr/. Released in February 2006, the most recent *QDR* reflects the Defense Department's focus on military transformation as it was emphasized by then-Secretary of Defense Donald Rumsfeld, as well as how ongoing military operations were reinforcing and affecting transformation efforts. Attributes of this transformation, as stressed in the *QDR*, include emphasizing that military threats have moved from reasonable predictability to an era of surprise and uncertainty; that threat planning must move from single-focused threats to multiple and complex challenges; the transition from nation-state threats to decentralized network threats from non-state enemies; the adjustment of conducting war against nations to conducting war in countries with which we are not at war; and the transition from one-size-fits-all deterrence to selectively customized deterrence for rogue governments, terrorist networks, and near-peer competitors.[13]

Additional examples of military force transformation heralded by the 2006 *QDR* include moving from major conventional combat operations to multiple irregular, asymmetric operations; stressing joint and combined operations instead of separate military service operational concepts; moving from set-piece maneuver and mass to agility and precision; transitioning from single-service acquisition systems to joint-portfolio management; transitioning from single-service and agency intelligence to Joint Information Operations Centers; moving from vertical structures and processes to more transparent and horizontal integration matrices; moving from static alliances to dynamic partnerships; and transitioning from static post-operations analysis to dynamic diagnostics and real-time lessons learned.[14]

The Joint Chiefs of Staff (JCS) (http://www.jcs.mil/) is also a major producer of U.S. military strategy documentation. The JCS Chair serves as the principal military advisor to the President, National Security Council, and Secretary of Defense. Other JCS staff members and professional staff provide advice on military matters to these individuals and organizations, including military strategic direction and planning, allocation of resources to fulfill such strategic plans, comparison of the capabilities of U.S. and allied armed forces with those of potential enemies, preparation and review of contingency plans conforming to presidential and Defense Department policy guidance, and preparation of other measures to ensure U.S. forces can implement the responsibilities they are given.[15]

The JCS has prepared a number of editions of *National Military Strategy of the United States* as assessments of U.S. military strategic objectives. The 1986 Goldwater-Nichols Reorganization Act gives the JCS Chair the responsibility of assisting the President and Secretary of Defense in providing strategic direction for the armed forces. The 1992 edition of this document stressed how the containment of the Soviet Union and communist ideology had been the primary focus of national military strategy in the previous decades. This document maintained that future threats to U.S. interests were derived from the uncertainty and instability of a quickly changing world and that a joint force of soldiers, sailors, airmen, and marines was essential to meet future security requirements.[16] Subsequent *National Military Strategy* reports were issued in 1995, 1997, and 2004. These documents reflect changes in U.S. military strategy over this long period, and they can be found through http://catalog.gpo.gov/ or other online resources. Topics addressed in the 2004 document include the role of national military strategy; the handling of a wider range of adversaries and a more complex battle space; agility, decisiveness, and integration as key strategic principles; U.S. military objectives, including protecting the United States, preventing conflict and surprise attacks, and prevailing against adversaries; the importance of having a joint military force with requisite capabilities to achieve mission success; and developing collaborative relationships with domestic and foreign partners.[17]

Additional pertinent Department of Defense (DOD) and JCS national military strategy documents include *National Defense Strategy of the United States* (2005), *National Military Strategic Plan for the War on Terrorism* (2006), and *National Military Strategy to Combat Weapons of Mass Destruction* (2006). Key characteristics of the first document, accessible at http://purl.access.gpo.gov/GPO/LPS59037, include securing the United States from direct attack by giving top priority to dissuading, deterring, and defeating those seeking to harm the United States directly with WMD; securing strategic access and retaining global freedom of action; strengthening alliances and partnerships; and establishing favorable security conditions. This document further states that these objectives will be implemented by developing active layered defenses, engaging in continuous transformation, developing a capabilities-based approach, and managing risks.[18]

The strategic plan for the war on terrorism is accessible at http://purl.access.gpo.gov/GPO/LPS66747. This document contends that the United States must confront a flexible and adaptable enemy in the Global War on Terror (GWOT):

> There is no monolithic enemy network with a single set of goals and objectives . . . In the GWOT, the primary enemy is a transnational movement of extremist organizations, networks, and individuals—and their state and non-state supporters—which have in common that they exploit Islam and use terrorism for ideological ends. The Al Qaida Associated Movement (AQAM), comprised of al Qaida and affiliated extremists, is the most dangerous present manifestation of such extremism. Certain other violent extremist groups also pose a serious and continuing threat.[19]

Critical military strategic objectives for fighting and winning a GWOT consist of denying terrorists what they need to operate and survive, such as mapping modes and connections, identifying the network, developing an action plan, tying the plan to metrics, and tracking progress to determine results. Additional components of this strategy include enabling partner nations to counter terrorism; denying WMD proliferation, recovering and eliminating uncontrolled materials, and increasing consequence management capacity; defeating terrorists and their organizations; and contributing to establishing conditions to counter ideological support for terrorism, including building security, providing humanitarian assistance, developing military-to-military contacts, conducting military operations in culturally sensitive ways, and developing information operations to assist moderate populations while countering extremist populations.[20]

National Military Strategy to Combat Weapons of Mass Destruction can be found at http://purl.access.gpo.gov/GPO/LPS68137. Critical components of the strategies detailed in this document include U.S. armed forces needing to carry out missions in the following areas: offensive operations, defensive operations, interdiction operations, active defense, passive defense, WMD consequence management, security cooperation and partnership activities, and threat reduction cooperation to prevent WMD detonations in U.S. territory.[21]

This document further identifies the six critical principles of U.S. strategy in this area as follows:

- Active, Layered, Defense-in-Depth
- Situational Awareness and Integrated Command and Control
- Global Force Management
- Capabilities-Based Planning
- Effects-Based Approach
- Assurance.[22]

Specific components of these principles include U.S. forces balancing, synchronizing, and coordinating all military WMD combating capabilities development and operations; having a highly flexible command and control process for dealing with actionable intelligence; being able to rapidly organize forces to conduct mission operations; developing tools that can be used in a wide variety of anti-WMD operations; and working effectively with international allies.[23]

Joint Doctrine Resources

The need for U.S. military forces to cooperate in conducting military operations has been noted by numerous political and military figures. In an April 3, 1958 address to Congress, President Eisenhower noted that separate ground, sea, and air warfare was gone forever and that future U.S. wars would involve all armed services branches and would require a single, concentrated effort to achieve success.[24] Later that year, with Eisenhower's advocacy, Congress would enact the Defense Reorganization Act, which began the long-term process of unifying military commands. This statute gave the President, acting in concert with the Secretary of Defense and with JCS advice, the authority to establish unified military commands, assign their missions, and determine their force structure. These military commands were correspondingly responsible to the Secretary and President for implementing their assigned missions. These commands were given full operational command over the armed forces assigned to them, which could only be transferred with presidential approval.[25]

In subsequent decades, major conflicts such as the Vietnam War, and smaller conflicts such as 1983's Operation Urgent Fury in Grenada saw the continuing presence of inter-service rivalry, which many critics of military organization saw as hampering military effectiveness. This criticism would ultimately result in the 1986 congressional passage of the Goldwater-Nichols Act, which sought to place increasing emphasis on joint service collaboration within the U.S. military. Although there are different assessments of the effectiveness of this act, the Goldwater-Nichols legislation put the ideal of joint collaboration between U.S. military services at the forefront of U.S. military policymaking and doctrinal development.[26]

This legislation gave operational military command authority to the Chair of the JCS instead of military service chiefs. The JCS Chair was designated the principal military advisor to the President, National Security Council, and Secretary of Defense. Goldwater-Nichols also established a Vice-Chair of the JCS, streamlined the operational chain of command from the President to the Secretary of Defense to the unified commanders, and served as the doctrinal basis for U.S. military operations in locales as scattered as Bosnia, Haiti, and the Persian Gulf region.[27]

Joint Electronic Library

Numerous doctrinal resources produced by the JCS in the two decades since the Goldwater-Nichols enactment have served to illustrate the critical role joint doctrinal thinking plays in U.S. military operations, planning, and policymaking. These resources are compiled in the Joint Electronic Library (JEL), which is accessible at www.dtic.mil/doctrine/. As of late September 2008, this library consisted of 77 joint doctrine publications covering a variety of topics. These publications are broken down into categories such as Capstone Publications, which includes *JP 1 Joint Doctrine for the Armed Forces of the United States* (2007), the most authoritative statement of U.S. joint military doctrine policy, and Reference Publications, which includes *JP 1–02 DOD Dictionary of Military and Associated Terms* (2001).

JEL also offers numerically arranged series of joint doctrine publications, including *JP 1 Joint Personnel Series, JP 2 Intelligence Series, JP 3 Joint Operations Series, JP 4 Logistics Series, JP 5 Joint Plans Series,* and *JP 6 C4 Systems Series.* Examples of some of the joint doctrine publications in these categories include *JP 1–04 Legal Support to Military Operations* (2007), *JP 2–01.3 Joint Tactics, Techniques, and Procedures for Joint Intelligence Preparation of the Battlespace* (2000), *JP 3–06 Joint Doctrine for Urban Operations* (2002), *JP 3–13.4 Military Deception* (2006), *JP 4–05 Joint Mobilization Planning* (2006), *JP 5–0 Joint Operation Planning* (2006), and *JP 6–0 Joint Communications System* (2006).[28]

These publications seek to detail U.S. joint military doctrine in all of the areas described. To provide a better understanding of how these documents are presented and arranged, a portion of their content will be reproduced here. This will help readers gain a heightened understanding of how these documents are organized and written.

JP 1 begins with an executive summary and chapter contents covering topics like U.S. military doctrine foundations; doctrine governing the unified direction of armed forces; functional characteristics of the DOD and major component organizations, including the JCS, military departments, and services; combatant commander responsibilities; joint doctrine command and control; doctrine for joint commands, including establishing unified and subordinate joint commands, discipline, and personnel administration; multinational operations, and interagency, intergovernmental organization, and nongovernmental organization coordination. Document appendices describe the role of doctrine and include relevant administrative instructions and a glossary of acronyms.[29]

This document describes the role of joint military operations in the following excerpt:

> The Joint Force. Twenty years after the Goldwater-Nichols Department of Defense (DOD) Reorganization Act…directed actions to remove the institutional barriers to jointness, the Armed Forces of the United States is a joint team. All Service components contribute their distinct capabilities to the joint campaign; however, their interdependence is critical to overall joint effectiveness. Joint interdependence is the purposeful reliance by one Service on another Service's capabilities to maximize complementary and reinforcing effects of both; the degree of interdependence varying with specific circumstances. Fundamentally, joint forces require high levels of interoperability and systems that are "born joint" (i.e., conceptualized and designed with joint architectures and acquisition strategies). This level of interoperability ensures that technical, doctrinal, and cultural barriers do not limit the ability of JFCs [Joint Force Commanders] to achieve objectives. The goal is to design joint force capabilities—lethal and nonlethal—to fight and win the Nation's wars and effectively carry out all other missions across the range of military operations.[30]

The complex domestic and international political and military requirements of early 21st-century military operations may require the U.S. military to interact with a variety of other civilian and military institutions. The following section of *JP 1* illustrates this situation:

> Complex operations, such as peace operations, may require a high order of civil-military integration. Presidential directives guide participation by all US civilian and military agencies in such operations. Military leaders must work with the other members of the national security team in the most skilled, tactful, and persistent ways to promote unity of effort. Operations of agencies representing the diplomatic, economic, and informational instruments of power are not under command of the Armed forces of the United States or any specific CCDR [combatant commander]. In domestic US situations, another department such as the Department of Homeland Security (DHS) may assume overall control of the interagency coordination including military elements. Abroad, the US ambassador and the country team may be in control in operations other than war not involving the use of force.[31]

The JP 3 series represents the most extensive collection of joint military doctrine publications. Focusing on operational activities, joint doctrinal topics addressed by these publications include shipboard helicopter operations, joint special operations, electronic warfare, psychological operations, command and control for joint maritime operations, joint engineer operations, and targeting.

JP 3–18 Joint Forcible Entry Operations (2008) covers issues addressed in this category of military operations, including principles for forcible entry operational success, forcible entry capabilities, command and control, planning, the purpose of such operations, integrating and synchronizing these operations, and logistics. An appendix covers amphibious assault operations and airborne and air assault operations.[32]

Key principles of successful forcible entry military operations, according to this document, include achieving surprise and gaining control of the contiguous air, spatial, and sea assets. More detailed instruction is provided as follows:

> Planners should try to **achieve surprise** regarding exact objectives, times, methods, and forces employed in forcible entry operations. The degree of surprise required depends on the nature of the operation to be conducted. Air superiority should be achieved in the operational area to **protect the force during periods of critical vulnerability** and to **preserve lines of communications.** At a minimum, the joint force must neutralize the enemy's offensive air and missile capability and air defenses to achieve local air superiority over the planned lodgment. **Space superiority** allows the joint force commander (JFC) access to communications, weather, navigation, timing, remote sensing, and intelligence assets without prohibitive interference by the opposing force. **Control of the sea** in the operational area enables the joint force to project power ashore in support of the joint forcible entry operation and to protect sea lines of communications.[33]

These JEL resources illustrate the richness of U.S. joint military doctrine, which is continuously revised and updated to accommodate changing military and political realities affecting the operational activities of U.S. military forces. JEL contains additional resources on doctrine beside the JP publications series. These include the text of research papers on joint doctrine, such as *U.S. Department of Defense Strategic Planning: The Missing Nexus* (1995), links to other U.S. military doctrinal service publications, articles analyzing U.S. military doctrine from the scholarly journal *Joint Force Quarterly* (1993–present), and research publications from the Joint Warfighting Center (JWFC) accessible at http://www.dtic.mil/doctrine/jwfc_pam.html.

Examples of JWFC publications include pamphlets such as *US Government Draft Planning Framework for Reconstruction, Stabilization, and Conflict Transformation* (2005), handbooks such as *Commander's Handbook for Joint Battle Damage Assessment* (2004) and *Joint Fires and Targeting Handbook* (2007), and white papers such as *Pre-Doctrinal Research White Paper No. 07–01 Provincial Reconstruction Teams* (2007).

Additional U.S. military sources providing analysis of joint U.S. military doctrine include *A Common Perspective Newsletter* from JWFC, United States Joint Forces Command (http://www.jfcom.mil/) and its component organizations, including JWFC and the Joint Warfare Analysis Center, Joint Forces Staff College (http://www.jfsc.ndu.edu/), various National Defense University (http://www.ndu.edu/) components, including the Institute for National Strategic Studies (http://www.ndu.edu/inss/), and Joint Special Operations University (https://jsoupublic.socom.mil/).

U.S. Air Force Doctrine Resources

U.S. Air Force doctrine has covered a multitude of subject areas during the Air Force's six-decade history as an independent U.S. armed service branch. Subject areas covered by Air Force doctrine include conventional aerial military operations such as bombing enemy targets, reconnaissance, attacking hostile air forces with fighter aircraft, supporting U.S. and allied ground forces in military operations, developing U.S. doctrine for using nuclear weapons through aerial bombers and intercontinental ballistic missiles, formulating U.S. military doctrine for conducting military operations in space and defending U.S. space assets against hostile military operations, and using aerospace power (a combination of aerial and space power) to fight counterinsurgency wars such as those currently ongoing in Afghanistan and Iraq. The history of U.S. military aerial doctrine has been shaped in various ways by individuals such as Henry A. "Hap" Arnold, Giulio Douhet, Ira Eaker, Laurence S. Kuter, Billy Mitchell, Carl Spaatz, and Hugh Trenchard. An extensive corpus of scholarly literature on the multiple factors driving the development of U.S. aerospace doctrine exists, producing sometimes contradictory assessments of the quality of this doctrine and its suitability for historical, current, or future U.S. military operations.[34]

Air Force Electronic Publishing

The Air Force's electronic publishing site (http://www.e-publishing.af.mil/) is the principal access point for Air Force policy documents, including service doctrinal resources. The annual Air Force Posture Statement, submitted to Congress as part of the Air Force's annual budget request, is accessible at http://www.posturestatement.af.mil/ and provides useful guidance for understanding current Air Force mission emphases and priorities. As of early October 2008, 32 Air Force Doctrine Documents (AFDD) are publicly accessible from Air Force Electronic Publishing and through the U.S. Government Printing Office's *Catalog of Government Publications* (http://catalog.gpo.gov/).

AFDD 1 Air Force Basic Doctrine (2003) is the capstone document explaining basic Air Force military doctrine principles. Examples of other Air Force military doctrine documents include *AFDD 2–1 Air Warfare* (2000), *AFDD 2–1.5 Nuclear Operations* (1998), *AFDD 2–1.7 Airspace Control in the Combat Zone* (2005), *AFDD 2–2 Space Operations* (2006), *AFDD 2–3 Irregular Warfare* (2007), and *AFDD 2–9 Intelligence, Surveillance, and Reconnaissance Operations* (2007).

AFDD 1 is divided into seven chapters whose contents cover an introduction to the nature of Air Force doctrine; the relationship between policy, strategy, doctrine, and war; aerospace power principles and tenets; aerospace power missions and functions such as strategic attack, counter-air, and combat support; expeditionary air force organization; core competencies and distinctive capabilities, including global attack and precision engagement; and linking future and present vision, operating concepts, and doctrine.[35]

AFDD 1 begins by describing attributes of good military doctrine, including doctrine being about war fighting instead of physics, effects not platforms, using mediums instead of owning mediums, synergy instead of segregation, integration as opposed to synchronization, and preserving national treasure. It also mentions that this particular Air Force doctrine was created by the Air Force Doctrine Working Committee, which is part of the Air Force headquarters Air Force Doctrine Center.[36] *AFDD 1* expresses war as a clash of opposing wills, describes changing characteristics in American war-fighting practice, and enumerates the unique attributes the Air Force brings to American military power:

> The US Air Force provides the Nation a unique capability to project national influence anywhere in the world on very short notice. Air and space forces, through their inherent speed, range, and flexibility, can respond to national requirements by delivering precise military power to create effects where and when needed. With expanding space and information capabilities, the Air Force is rapidly developing the ability to place an "information umbrella" over friends and foes alike. This provides national political and military leaders with unprecedented knowledge of world events; fosters rapid, accurate military decisions; and directly complements the Service's air and space power forces, while at the same time denying potential adversaries access to useful information on our own plans, forces, and actions. The US Air Force,

> in fielding advanced, highly effective, lethal and nonlethal systems, provides national leaders and joint force commanders (JFCs) unique capabilities across the range of military operations.[37]

Precision engagement is also a critical attribute of U.S. Air Force military doctrine as this excerpted passage from *AFDD 1* shows:

> Increasingly, air and space power is providing the "scalpel" of joint Service operations—the ability to apply discriminate force precisely where required. Precision engagement is the ability to command, control, and employ forces to cause specific, strategic, operational, or tactical effects. The Air Force is clearly...the Service with the greatest capacity to apply the technology and techniques of precision engagement anywhere on the face of the Earth in a matter of hours. In addition to the traditional application of force, precision engagement includes nonlethal as well as lethal force. Functions such as the close surveillance of peace agreements between belligerents by airborne and space-based assets, the employment of AFSOF [Air Force Special Operations Forces] in small-scale but precise operations, or the rapid response of airlift to the source of an erupting humanitarian disaster are prime examples of precision engagement. Precision engagement represents a global capability not only to win wars, but also the ability to drive crises to peace.[38]

The Air Force has played a key role in developing U.S. nuclear weapons strategic doctrine due to its responsibilities for the air component of the military's nuclear weapons triad. *AFDD 2–1.5 Nuclear Operations* serves as the Air Force's authoritative documentation of U.S. nuclear operations strategy if wartime conditions require unleashing the United States' nuclear arsenal. The first chapter discusses the roles played by deterrence in nuclear operations emphasizing ICBMs, bombers, theater-range weapons, and the safety and security of nuclear weapons systems. Chapter two discusses nuclear weapons command and control, including authorization for nuclear weapons use; weapons system safety rules; communication system survivability and redundancy; and Air Force organization for continental U.S.-based nuclear operations. Chapter three examines planning and support considerations, such as logistics, and chapter four examines the importance of training to ensure readiness and preparedness.[39]

AFDD 2–1.5 begins by acknowledging that the presence of significant Russian and Chinese nuclear weapons capabilities could threaten the United States, while also emphasizing that new threats could emerge from unknown sources. Consequently, the United States requires a nuclear weapons deterrent because it does not have the ability to respond to chemical or biological weapons attacks against it.[40] The critical importance of the concept of nuclear deterrence to U.S. military strategy and doctrine is described in the following passage:

> *Deterrence can be described as a state of mind in an adversary's (or potential adversary's) leadership.* Their leadership must believe the cost of aggression against the United States, its interests, or its allies will be so high as to outweigh any

> possible gain. Deterrence requires the United States to maintain the ability to use force, which means having trained capable, ready, and survivable forces; a robust command, control, communications, computers and intelligence structure; and timely, flexible, and adaptive planning capabilities. The second critical element of deterrence is the will to use nuclear weapons. If an enemy believes these tools will not be used, their deterrent value is zero.[41]

U.S. nuclear weapons use doctrine may involve counter-value targeting and counterforce strategy. Counter-value targeting consists of holding enemy cities, industry, and other economic resources at risk by striking critical infrastructures or primary production means, including harbors, industrial centers, or oil pipelines. Counterforce strategy involves using weapons against an enemy's primary war fighting capabilities, which may involve destroying hostile WMD forces before they can be used, or using weapons against an adversary's conventional forces if U.S. or allied conventional warfare has proven unsuccessful. Such strategy can reduce the threat to the United States and its force, destroy enemy forces, and result in conflict termination.[42]

The Lemay Center for Doctrine Development and Education at Maxwell-Gunter Air Force Base in Montgomery, AL has served as the Air Force's center for education, war gaming, and doctrine development since August 2, 2007.[43] The Lemay Center's Web site (http://www.cadre.maxwell.af.mil/) provides links to a variety of doctrinal resources and analysis of Air Force doctrine produced by Air University, which serves as the Air Force's professional military educational institution. These include *Air and Space Power Journal* (http://www.airpower.au.af.mil/) and its predecessors, *Aerospace Power Journal* and *Air University Review,* which have been published since 1947, and the new journal, *Strategic Studies Quarterly* (2007–present).

Analysis of Air Force doctrinal history and development is also published by Air University Press (http://aupress.maxwell.af.mil/), which includes the full text of many of its books and monographic series such as CADRE papers. Theses from students at Air University's School of Advanced Airpower Studies and other schools can be found at https://research.au.af.mil/showstudent.aspx?type=student. These documents provide insights into Air Force doctrinal issues from emerging officers.

Additional assessments of Air Force doctrinal issues are produced by the Air Force Academy's Institute for National Security Studies (INSS) (http://www.usafa.af.mil/df/inss/). INSS's Occasional Papers series provides access to 66 analyses of national security policy issues, including Air Force doctrine from 1994 to the present.

The Rand Corporation is a major national security-oriented public policy research institution that has done contractual work for the Air Force and other military services for several decades. Its Project Air Force (PAF) (http://www.rand.org/paf/) provides numerous publicly available analyses of Air Force military and doctrinal issues and includes a Strategy and Doctrine division. Examples of analyses of Air Force doctrine produced by PAF include *Future Roles of U.S. Nuclear*

Forces: Implications for U.S. Nuclear Strategy (2003), *Striking First: Preemptive and Preventive Attack in U.S. National Security Policy* (2006), *Learning Large Lessons: The Evolving Role of Ground and Air Power in the Post-Cold War Era* (2006), and *Dangerous Thresholds: Managing Escalation in the 21st Century* (2008). PAF also maintains an active research agenda with its Strategy and Doctrine divisions. Its 2008 research agenda includes topics such as potential Air Force operational roles in Iraq once U.S. forces are drawn down, future requirements and options for U.S. nuclear forces, assessment of Air Force security cooperation activities with other countries, counters to Chinese military space power, and evaluation of Air Force force structure for major combat operations.[44]

Army Doctrine Resources

The U.S. Army has been in existence for over two hundred years and Army doctrine for conducting military operations has been continually updated. Numerous U.S. and foreign military figures have influenced U.S. military doctrinal development, including Carl von Clausewitz, William DuPuy, Antoine-Henri Jomini, Basil Liddell-Hart, David Petraeus, Emory Upton, and many others. U.S. army doctrine encompasses a wide variety of land force operations as well as the coordination of these operations with aerospace and naval forces. Examples of topics addressed by U.S. Army doctrine documents include intelligence, special operations forces, logistics, detainee operations, military police activities, operating in biological, chemical, and nuclear battlefield environments, casualty treatment and battlefield evacuation, counterinsurgency, peacekeeping, various forms of humanitarian assistance, and numerous other topics. Recent emphases of Army military doctrine have focused on the complexities of conducting counterinsurgency operations in Afghanistan and Iraq, while striving to win the support of indigenous populations in those countries. There is an extensive and proliferating literature on the historical successes, failures, lessons learned, and uncertainty of U.S. Army doctrine, with speculation of how this doctrine may or may not succeed in meeting current and future U.S. political, diplomatic, and military objectives in conflicts around the world requiring U.S. military intervention.[45]

General Dennis J. Reimer Training and Doctrine Digital Library

The Reimer Library (https://rdl.train.army.mil/) is named after the general who was U.S. Army Chief of Staff from 1995–1999. This resource provides access to a lot of Army training resources, including Field Manuals (FM), which are the most important sources of army doctrinal information. Some FMs are classified, but a September 2008 search of 491 FMs in the Reimer Training and Doctrine Library found 399 (81.2%) accessible to the general public. The annual Army Posture Statement (http://www.army.mil/aps/) also provides useful information on current Army mission objectives and planning. Many U.S. Army FMs can also be found on the Web site of the research organization globalsecurity.org at http://www.globalsecurity.org/military/library/policy/army/fm/.[46]

FMs are numbered sequentially and provide detailed guidance as to how Army units and personnel are to conduct various kinds of military operations. *FM 1 The Army: Our Army at War Relevant and Ready Today and Tomorrow* (2005) serves as the overall theoretical guidance for service doctrine. It mentions that the Army is ready to address traditional, irregular, catastrophic, and disruptive security challenges that may require it to defend the United States. *FM 1* goes on to mention that the Army is seeking to prevail in major combat operations by enhancing its capabilities in the following areas:

- Strategic and operational mobility
- Advanced information systems to support command, control, intelligence, surveillance, and reconnaissance
- Precision weaponry
- Force protection
- Sustainment[47]

This document also asserts that the Army is enhancing its ability to counter irregular challenges by increasing the versatility and agility of forces conducting conventional operations; preempting catastrophic threats, such as deterring the use of or destroying mass destruction weapons; increasing its ability to rapidly project and decisively maneuver forces on both global and theater distances; and seeking minimal reliance on predictable and vulnerable transition points, such as staging bases or ports of entry.[48]

Examples of specific, publicly accessible Army FMs, which are revised and updated on an ongoing basis, include *FM 1–100 Army Aviation Operations* (1997), *FM 3–01.16 Procedures for Theater Missile Defense Intelligence Preparation of the Battlespace* (2002), *FM 3–22.9 Rifle Marksmanship M16A1, M16A2/3, M16A4 and M4 Carbine* (2006), *FM 3–24 Counterinsurgency* (2006), *FM 4–02.51 Combat and Operational Stress Control* (2006), *FM 5–71–3 Brigade Engineer Combat Operations (Armored)* (1997), and *FM 6–20 Fire Support in the Airland Battle* (1988).

FM 3–24, which covers counterinsurgency operations, should be of particular interest given ongoing U.S. military operations in Afghanistan and Iraq, which have proven to be excellent test beds for revising and refining U.S. counterinsurgency doctrine. Topical themes addressed in *FM 3–24* include integrating civilian and military activities, the importance of intelligence like battlefield planning and preparation, protecting sources, developing host nation security forces, the importance of ethical conduct toward indigenous populations, distinguishing between war fighting and policing, selecting qualified and loyal interpreters, developing effective legal detention and interrogation practices, enforcing discipline of U.S. forces, and providing humanitarian relief and reconstruction.[49]

The following excerpt from *FM 3–24* describes the important interrelationship between war fighting and policing and the critical ethical importance of military and civilian forces working together to achieve desired political and military objectives:

> In counterinsurgencies, warfighting and policing are dynamically linked. The moral purpose of combat operations is to secure peace. The moral purpose of policing is to maintain the peace. In COIN [counterinsurgency] operations, military forces defeat enemies to establish civil security; then, having done so, these same forces preserve it until host-nation (HN) police forces can assume responsibility for maintaining the civil order. When combatants conduct stability operations in a way that undermines civil security, they undermine the moral and practical purposes they serve. There is a clear difference between warfighting and policing. COIN operations require that every unit be adept at both and capable of moving rapidly between one and the other.[50]

The vital imperative of securing and holding acquired territory in counterinsurgency operations is reflected in the following *FM 3–24* analysis:

> The COIN environment frequently and rapidly shifts from warfighting to policing and back again. There are many examples from Iraq and Afghanistan where U.S. forces drove insurgents out of urban areas only to have the insurgents later return and reestablish operations. Insurgents were able to return because U.S. forces had difficulty maintaining civil security. U.S. forces then had to deal with insurgents as an organized combatant force all over again. To prevent such situations, counterinsurgents that establish civil security need to be prepared to maintain it. Maintaining civil security entails very different ethical obligations than establishing it.[51]

U.S. Army Training and Doctrine Command

The Army's Training and Doctrine Command (TRADOC) (http://www.tradoc.army.mil/) is also a critical U.S. Army doctrinal information resource. Established in 1973 and headquartered at Fort Monroe, VA, TRADOC is responsible for recruiting, training, and educating Army soldiers; developing leaders; and developing Army doctrine, including field manuals, which describe how the Army fights tactically and how tactics and weapons systems are integrated into Army operations.[52]

TRADOC mission activities are carried out by component organizations such as the Army Capabilities Integration Center (ARCIC) (http://www.arcic.army.mil/), the Combined Arms Center (CAC) at Fort Leavenworth, KS (http://uscac.army.mil/CAC2/), and Combined Arms Support Command (CASCOM) at Fort Lee, VA (http://www.cascom.lee.army.mil/). ARCIC is responsible for identifying, designing, developing, and synchronizing capabilities into the Army's current and future modular force structures, while supporting TRADOC to provide adaptive soldiers, leaders, and units into doctrine development. CAC is responsible for preparing the Army and its leaders for war, focusing such preparation on fighting terrorism and meeting future conventional threats. CASCOM is responsible for providing training and leader development and developing doctrine organizations and educational and material support to sustain a campaign-quality Army with joint and expeditionary force capabilities.[53]

Strategic Studies Institute

The U.S. Army War College's Strategic Studies Institute (SSI) (http://www.strategicstudiesinstitute.army.mil/) is another important resource for analyzing U.S. Army doctrine. SSI serves as the Army's geostrategic and national security research and analysis institute. Its work supports Army War College curricula, provides analysis for Army and DOD leadership, and serves as a conduit for interaction with the broader security studies community in governments, militaries, and academe. Its personnel include civilian research professors, uniformed military officers, and a professional support staff. SSI component entities include the Strategic Research and Analysis Department, which focuses on global, transregional, and functional issues, such as doctrine, and a Regional Strategy and Planning Department, which emphasizes regional strategic matters.[54]

Examples of SSI analyses of Army doctrine include *The Owl of Minerva Flies at Twilight: Doctrinal Change and Continuity and the Revolution in Military Affairs* (1994), *Problems and Solutions in Future Coalition Operations* (1997), *The Interagency and Counterinsurgency Warfare: Stability, Security, Transition, and Reconstruction Roles* (2007), *U.S. Army War College Guide to National Security Issues Volume I: Theory of War and Strategy*, 3rd ed. (2008), and *Stability Operations and State-Building: Continuities and Contingencies* (2008).

The Army War College's scholarly journal, *Parameters* (http://purl.access.gpo.gov/GPO/LPS1511), is also an excellent resource for analysis and debate on Army and other military doctrinal issues.

Combat Studies Institute

The Combat Studies Institute (CSI) (http://usacac.army.mil/cac2/csi/) is part of the Army's Combined Arms Center (CAC) and Command and General Staff College (CGSC) at Fort Leavenworth, KS. CSI's mission is providing timely and relevant military history research publications and contemporary operational history for the Army.[55]

CSI's publishing division (CSI Press) provides access to a wide variety of analyses of historical and contemporary Army doctrinal issues. Examples of these publications include *On Point: The United States Army in Operation Iraqi Freedom* (2004), *Field Artillery in Military Operations Other Than War: An Overview of the U.S. Experience* (2004), *Boots on the Ground: Troop Density in Contingency Operations* (2006), and *We Were Caught Unprepared: The 2006 Hezbollah-Israeli War* (2008), which are part of CSI's Long War Occasional Papers monographic series.

Additional CSI Press resources include Leavenworth Papers monographic series titles, such as *The Dynamics of Doctrine: The Changes in German Tactical Doctrine During the First World War* (1981), and other reports and masters-level theses, including *Sixty Years of Reorganizing for Combat: A Historical Trend Analysis*

(1999), *Adequacy of Current Interagency Doctrine* (2007), *Adopting a Single Planning Model at the Operational Level of War* (2008), *Creating Effective Post-Conflict Transition Organizations: Lessons from Panama, Bosnia, Afghanistan, and Iraq* (2008), and *Exploitation Tactics: A Doctrine for the 21st Century* (2008).

The scholarly journal *Military Review* (http://purl.access.gpo.gov/GPO/LPS53409) is also an excellent source for analysis of army and other military doctrinal matters.

An additional noteworthy resource for Army doctrine discussion and analysis is CAC's Blog Library (http://usacac.army.mil/blog/). This forum features postings and comments from participants on a wide variety of military policy issues, including Army doctrine. Examples of topics addressed and discussed are stability operations doctrine, transitions while conducting counterinsurgency operations, and updating the new army training manual *FM 7–0*.

CAC and CGSC students are current and emerging Army leaders whose career trajectories may put them in positions to write future Army doctrine documents. Using CSI resources produced by these students and other individuals is an excellent way to determine and assess potential future directions in U.S. Army doctrinal thinking.

Association of the U.S. Army Institute of Land Warfare

The Association of the United States Army (AUSA) is a private non-profit educational organization founded in 1950 to support the U.S. Army, reserves, civilian army employees, and their families.[56] AUSA's Institute of Land Warfare (ILW) (http://www.ausa.org/about/ilw/) seeks to educate its members, governmental leaders, and the general public about the vital importance of land forces and the U.S. Army by publishing a variety of reports and information resources on these topics.[57]

Examples of publications produced by ILW include its Background Briefs, Defense Reports, Land Warfare Papers, and Land Power essays, which analyze trends and developments affecting military land forces. Representative examples of ILW publications examining Army doctrine include *Gun-Fired Precision Munitions for a Transformed Army* (2003), *Surprise, Shock, and Daring: The Future of Mobile, All-Arms Warfare* (2004), *Defining Asymmetric Warfare* (2006), *Implications of Laser Weapons for Ground Combat Operations* (2006), *Planning for the Employment of the Reserve Components: Army Practice, Past and Present* (2008), and *Tactics for Small Wars* (2008).

Rand Arroyo Center

The Rand Corporation's Arroyo Center (http://www.rand.org/ard/) was founded in 1982 as NASA's Jet Propulsion Laboratory and was moved to Rand in 1984 at the request of the Army's Chief of Staff. The Arroyo Center serves as the Army's only federally funded research and development center for studies and analysis. Its research programs cover strategy, doctrine, and resources, including how a changing security environment may affect future Army roles structure and doctrine; force development and technology assessing technological advances and emerging operational concepts to enhance Army mission performance; military logistics to improve Army operational force support and industrial base and support infrastructure; and manpower and training using economic and social science methodologies to enhance Army personnel quality and training. Additional Arroyo Center research emphases include:

- Conducting objective analytic research on major policy matters emphasizing mid- to long-term policy issues;
- Helping the Army improve its effectiveness and efficiency;
- Providing short-term assistance on urgent problems; and
- Serving as a catalyst for needed change.[58]

Examples of Arroyo Center resources analyzing Army doctrine include *National Security Newsletter to Congress* (2002–present), *Army Futures and the Army Force Plan: Implications for the Future Force Era* (2005), *Army Forces for Sustained Operations* (2005), *The Impact of Network Performance on Warfighter Effectiveness* (2006), *Preparing the Army for Stability Operations: Doctrinal and Interagency Issues* (2007), *Green Warriors: Army Environmental Considerations for Contingency Operations from Planning Through Post–Conflict* (2008).

United States Marine Corps Doctrine Resources

The United States Marine Corps (USMC) has developed its own unique corpus of doctrinal literature during its historical and contemporary development and evolution. This literature has emphasized unique aspects of Marine service operational thinking, such as stressing the importance of ship-to-shore amphibious operations and being the first U.S. military service to stress the importance of counterinsurgency operations and fighting small wars as part of its military doctrine. Individuals such as Sir Julian Corbett, Alfred Cunningham, Archibald Henderson, Thomas Holcomb, and John Lejeune have had significant influence on Corps organizational structure and doctrinal thinking, with the 1940 *Small Wars Manual* being a particularly significant work with continuing relevance. Numerous assessments of USMC doctrine have been published by Marine and non-Marine authors in a variety of forums.[59]

Marine Corps annual posture statements to Congress, as part of its annual budget justification requests, are good information sources for examining contemporary USMC thinking on overall operational issues. The most recent Corps posture statement can be found through the USMC Commandant's Web site (http://www.marines.mil/units/hqmc/cmc/). The official Marine Corps doctrinal site (https://www.doctrine.quantico.usmc.mil/) is not accessible to the general public.

However, there are other options for accessing USMC doctrinal resources. Some resources are accessible through JEL, and others are available through the Corps Orders and Directives: Doctrine Pubs Web site (http://www.marines.mil/news/

publications/Pages/order_type_doctrine.aspx). A listing of these documents is accessible through the USMC Artillery Detachment at Fort Sill, OK (http://sill-www.army.mil/USMC/Pubs/).

There are different categories of USMC doctrine publications. Marine Corps Doctrinal Publications (MCDP) are higher order doctrine publications containing foundational and enduring war-fighting beliefs. They are broken up into Capstone Publications, such as *MCDP 1 Warfighting* (1997) and *MCDP 1–2 Campaigning* (1997), and Keystone Publications classified into the *MCDP 2–6* series, with representative samples including *MCDP 2 Intelligence* (1997), *MCDP 3 Expeditionary Operations* (1998), *MCDP 4 Logistics* (1997), *MCDP 5 Planning* (1997) and *MCDP Command and Control* (1996). Marine Corps Warfighting/Reference Publications (MCWPs/MCRPs) are more specifically focused on detailing tactics, techniques, and procedures used by the Corps to prosecute war and other assigned tasks. Examples of some of these publications include *MCWP 2–14 Counterintelligence* (2000), *MCWP 3–11.3 Scouting and Patrolling* (2000), and *MCRP 3–16C Tactics, Techniques, and Procedures for Fire Support for the Combined Arms Commander (2001)*. Additionally, Fleet Marine Force Manuals (FMFM) such as *FMFM 3–3 Helicopterborne Operations* (1972) provide operational guidance for conducting combat operations and are accessible at http://www.marines.cc/content/view/82/56/.[60]

Since expeditionary operations and projecting military power from ship to shore in the form of amphibious assaults have been hallmark characteristics of USMC operational activities, consulting *MCDP 3* can be particularly instructive for understanding Corps operational thinking. This treatise stresses that expeditionary warfare refers to austere conditions and support levels, which means that such forces are only equipped with the supplies and infrastructure to meet operational necessities. Expeditionary bases or airfields used to carry out operational missions are given less than the usual range of support associated with permanent stations. Force protection and intelligence take precedence over administrative, quality of life, and other considerations. This insistence on austerity stems from security considerations, the temporary nature of expeditionary operations, and the criticality of minimizing lift and support requirements.[61]

This emphasis on operational agility and minimizing stationary activity is a critical characteristic of the Corps's expeditionary warfare doctrine. This is contrasted with the practices of other U.S. armed services where expeditionary operations are concerned:

> to perform expeditionary operations requires a special mindset—one that is constantly prepared for immediate deployment overseas into austere operating environments, bringing everything necessary to accomplish the mission . . . In general, naval expeditionary forces provide a self-sustaining, sea-based capability for immediate or rapid response, especially through forward deployment. Land-based forces, on the other hand, generally require a longer deployment phase and the creation of an in-theater logistics apparatus to achieve

> the buildup of decisive force. While all the Services include units capable of expeditionary operations, the entire operating forces of the Marine Corps are specifically organized, equipped, and trained for expeditionary service.[62]

Accurate intelligence gathering and analysis is critical to the success of any military operation regardless of which service branch conducts that operation. *MCWP 2–14 Counterintelligence* serves as the USMC's guide for conducting counterintelligence (CI) operations. This work mentions that operations, investigations, collection and reporting, and analysis, production, and dissemination are the four primary CI functions. The objectives of CI operations are determining foreign intentions; supporting tactical and strategic perception management operations; supporting all-source intelligence and other CI operations; and supporting planning and military operations. CI investigation attributes include detecting, exploiting, preventing, or neutralizing espionage activities; detecting and resolving foreign-directed sabotage, subversion, sedition, terrorist activities, and assassinations; documenting proof of such events for prosecution; and providing military commanders and policymakers with intelligence that can be used to eliminate security vulnerabilities and improve overall security.[63]

CI collection and reporting characteristics include providing indications and warning of security threats to U.S. forces, facilities, and operations; providing intelligence on threats to U.S. forces, facilities, and operations; providing intelligence on threats to forces to support planning and implementation of defensive or offensive countermeasures; and responding to commanders' priority intelligence requirements. CI analysis, production, and dissemination involves providing analysis and assessments of threats to U.S. forces, facilities, and operations; providing causal analysis of past events to identify emerging vulnerabilities and threats; and identifying adversary organizations, personalities, and capabilities that may threaten forces, facilities, and operations.[64]

Marine Corps Combat Development Command

The Marine Corps Combat Development Command (MCCDC) (https://www.mccdc.usmc.mil/) is responsible for developing completely integrated Corps capabilities, including doctrine, organization, training, education, and other assets, to enable the deployment of combat-ready forces.[65] An important MCCDC organizational component is the Operations Analysis Division, which studies and analyzes the Corps's combat development process to assist in making combat development decisions and applications to war-fighting capabilities.[66]

Additional Marine Corps Doctrinal Resources

Supplemental USMC entities that analyze doctrine and other operational issues include the Marine Corps Warfighting Laboratory's Small Wars Center of Excellence (http://www.smallwars.quantico.usmc.mil/), whose institutional objective is to understand the history and challenges of the Corps's involvement in small wars,[67] the Center for Advanced Operational Cultural Learning (http://www.tecom.usmc.mil/caocl/), which seeks to ensure that Marines have operationally pertinent regional, cultural, and language knowledge to allow them to operate successfully in joint and combined expeditionary environments in any global region,[68] and Marine Corps University (http://www.mcu.usmc.mil/), which serves as the Corps's professional military educational institution to develop skilled wartime leaders capable of critical and sound decisionmaking.[69]

The Marine Corps University Library (http://www.mcu.usmc.mil/MCRCWeb/) features access to university student papers analyzing doctrinal and other issues from 1984 to the present. Examples of some of these papers are "Air Land Battle and Maneuver Warfare: Do We Need Both?" (1989), "Amphibious Warfare and the Composite Warfare Commander" (1992), "World War II USMC and Navy Amphibious Doctrine: A Sound Set of Principles for the Time" (1999), "The Applicability of Maneuver Warfare to Counterinsurgency Operations" (2005), and "Urban Breaching Doctrine: Repairing the Cracked Foundation" (2006). Since the authors of these papers are likely to become future U.S. and foreign Marine leaders, their writings can provide some insight into how they approach military doctrinal issues.

Marine Corps University Press (http://www.tecom.usmc.mil/mcupress/) is also beginning to serve as a forum for disseminating Corps doctrine analysis. *Operational Culture for the Warfighter: Principles and Applications* (2008) and *Among the People: U.S. Marines in Iraq* (2008) are two relevant books it has already published, and *U.S. Marines and Irregular Warfare, 1898–2007: Anthology and Selected Bibliography* is slated for publication in 2008. In addition, this publisher will introduce the scholarly, multidisciplinary *Marine Corps University Journal* in mid–2009. This journal will become biannual in 2010 and will undoubtedly be a useful tool for analyzing Marine Corps doctrine.[70]

Further analysis of Marine Corps doctrine can be found in numerous military and strategic studies journals, including the *Marine Corps Gazette,* published by the Marine Corps Association. General information about this journal is available at http://www.mca-marines.org/. Additional analysis may be found in scholarly military history monographic literature.

United States Navy Doctrine Resources

United States Navy doctrine has been influenced by a number of individuals, including Philip Colomb, Julian Corbett, Dudley Knox, Stephen Luce, Alfred T. Mahan, Chester Nimitz, and Hyman Rickover. Throughout its existence, the Navy has grown from a small coastal protection force to the world's preeminent naval power with global reach and striking power. U.S. Navy doctrine has covered areas such as the importance of maintaining open international sea lanes and lines of communication; conventional naval operations such as combat between warships like battleships and cruisers; submarine warfare; naval aviation, including the power projection capabilities of aircraft carriers and the use of nuclear weapons through submarine-launched ballistic missiles. This doctrinal literature also focuses on combating piracy and conducting operations in littoral bodies of water, such as in areas adjacent to shorelines and rivers flowing into oceans. A continually growing body of knowledge of naval doctrine is accessible to interested students and scholars.[71]

Annual Navy posture statements are useful indicators of current service thinking on operational and strategic issues. The three most recent Navy posture statements are accessible through the Secretary of the Navy's Web site at http://www.navy.mil/navydata/leadership/ldrDisplay.asp?m=325. The October 2007 *A Comparative Strategy for 21st Century Seapower* (http://www.navy.mil/maritime/MaritimeStrategy.pdf) is the Navy's most recent strategic planning document.

Current Navy Doctrinal Publications (NDP) are accessible through the Joint Electronic Library. These resources are *NDP 1 Naval Warfare* (1994), *NDP 2 Naval Intelligence* (n.d.), *NDP 4 Naval Logistics* (2001), *NDP 5 Naval Planning* (n.d.), and *NDP 6 Naval Command and Control* (1995). Topics addressed in *NDP 1* include the nature of naval services and the character of naval forces; the employment of naval forces, emphasizing the roles played by forward presence, naval operations other than war, sealift, joint operations, and wartimes naval operations; how the Navy fights; and where the Navy is headed in the future.[72]

NDP 1 begins by stressing that the U.S. is a maritime nation with multiple interests, including global economic interdependence and a heritage intimately interwoven with its geographic location. It acknowledges that intercontinental commercial flights and instantaneous global communications have allowed new trade opportunities and brought nations closer together, while recognizing that we still rely on oceans for defense purposes and to serve as a global trade gateway. *NDP 1* stresses that 90 percent of the world's trade and 99 percent of U.S. import-export tonnage is transported by sea and that the U.S. economy is not self-sufficient as it remains dependent on the continuing flow of raw materials and finished products and services to and from the United States. Consequently, *NDP 1* declares that "ensuring that world's sea lanes remain open is not only vital to our own economic survival; it is a global necessity."[73]

This document proceeds to mention that naval forces have been organized to fight at sea for over two millennia, and the following passage describes the most critical attributes of modern U.S. naval forces:

> These qualities are *readiness, flexibility, self-sustainability,* and *mobility.* They permit naval forces to be expeditionary—that is, being able to establish and maintain a forward-based stabilizing presence around the world. Naval expeditionary operations are offensive in nature, mounted by highly trained and well-equipped integrated task forces of the Navy and Marine Corps organized to accomplish specific objectives. Naval expeditionary forces draw upon their readiness, flexibility, self-sustainability, and mobility to provide the National Command Authorities the tools they need to safeguard such vital national interests as the continued availability of oil from world producers and

> maintenance of political and economic stability around the globe. Through these qualities, naval forces reassure allies and friends, deter aggressors, and influence uncommitted and unstable regimes.[74]

NDP 1 also stresses the paramount importance of mobility in initiating and sustaining naval operations, as this excerpt demonstrates:

> Mobility is the key to decisive naval operations. The ability to maneuver ships into position to strike vulnerable targets, or to threaten amphibious assault at multiple locations along an extended coastline, is a significant tactical and operational advantage. After we have launched our strikes, our ships can press the advantage, maneuver out of range, or reposition themselves for the next strike phase. In amphibious operations, we place troops in a position to attack the weakness of the enemy while avoiding his main strength. A landing force's ability to maneuver from attack positions over the horizon through designated penetration points—without a slowdown or loss of momentum—could be critical to the success of the landing. When the Marines have accomplished their mission ashore, they can backload to await the next contingency.[75]

NDP 6 provides detailed elucidation of the importance of command and control in naval operations. One section of this document emphasizes the importance of observation, orientation, decision, and action (known as the OODA Loop) in the leadership and execution decision-making cycle. This process begins when a commander *observes* the environment using sensors, information systems, and situation reports from subordinates to collect data about his surroundings and the status of allied and hostile forces. Acquired data are then sorted, fused, and displayed together to present a common tactical picture of the existing battle space, which is then shared with other commanders. This intelligence process continues as the commander *orients* himself to the environment by forming a mental picture of the situation and converting sensor data and other information into estimates, assumptions, and judgments about what is occurring. Such orientation enables the commander to *decide* on a course of action, which he does by announcing his intent and issuing orders to take *action*. This action involves the commander monitoring operational executions and measuring their results, which results in a return to the OODA cycle. It must be emphasized that friction and the fog of war may continually hinder the commander's OODA capabilities.[76]

Additional attributes of naval command and control include Navy and Marine Corps forces being tailored for joint operations and scaled to missions, being organized in a way in which structural authority and responsibility are clearly defined, and making every organizational decision a command and control decision. The following passage indicates the importance of interconnected relationships at all levels of the chain of command:

> Organization establishes the *chain of command* and the command and support relationships within the force. The chain of command establishes authority

> and responsibility in an unbroken succession. Commanders at each echelon respond to intent and orders to their subordinates; each commander has full authority and responsibility within their given sphere. Command and support relationships specify the type and degree of authority one commander has over another and the type and degree of support that one commander must provide another.[77]

Navy Warfare Development Command

The Navy Warfare Development Command (NWDC) (http://www.nwdc.navy.mil/) is located in Norfolk, VA and Newport, RI. NWDC's responsibilities include developing concepts and doctrine to enable the Navy to enhance its maritime operational capacity and cooperate effectively with other U.S. armed services and coalition partners.[78]

Although some sections of NWDC's Web site are restricted to .mil users, useful information about navy doctrine can be gleaned here. This includes description of *Sea Power 21,* which is the operational basis for Navy doctrinal strategy in the 21st century. *Sea Power 21* emphasizes several concepts, including Sea Shield, Sea Strike, Sea Basing, Sea Warrior, Sea Trial, Sea Enterprise, and FORCEnet. Sea Shield seeks to develop naval capabilities pertaining to homeland defense, sea control, assured access, and overland defense projection. Sea Strike emphasizes augmented naval power projection through C4ISR, precision, stealth, and endurance to increase operational tempo, reach, and effectiveness.

Sea Basing projects U.S. sovereignty globally, while giving Joint Force commanders critical sea-based command and control, fire support, and logistics and minimizing vulnerable shore-borne assets. Sea Warrior strives to enhance the education and training process for developing 21st-century sailors. Sea Trial is an ongoing conceptual and technology development process emphasizing focused war games, experiments, and exercises to augment naval innovation culture and deliver enhanced capabilities to the fleet. Sea Enterprise captures efficiencies by employing lessons learned from the business world to target areas for improvement and prioritized resource allocation. FORCEnet seeks to integrate warriors, sensors, networks, command and control, platforms, and weapons into a network-centric combat force enabling network-centric warfare.[79]

NWDC's Web site also includes a Lessons Learned section that features hilarious Windows Media videos of a talking pirate skull named Captain Moby, who describes prominent historical Navy operations. It also includes recent historical strategy documents, such as *From the Sea: Preparing the Naval Service for the 21st Century* (1992) and *Forward From the Sea* (1994).

Chief of Naval Operations

The Chief of Naval Operations (CNO) is the Navy Department's senior military officer. This individual is a four-star admiral responsible to the Secretary of the Navy who advises that official on command, resource utilization, and Navy operating efficiency. The CNO is a JCS member and the principal naval advisor to the President and Secretary of the Navy.[80]

The CNO's Web site (http://www.navy.mil/navydata/leadership/ldrdisplay.asp?m=11) provides additional information about this office's responsibilities, including interviews and some historic Navy posture statements.

Naval War College

The U.S. Naval War College (NWC) (http://www.nwc.navy.mil/) is located in Newport, RI and serves as the Navy's principal professional military educational institution. Throughout its existence, NWC has sought to develop the Navy as it carries out its roles and missions. It promotes the development of naval officers and cooperation with allied navies through the Naval Command College and Naval Staff College. NWC's Center for Naval Warfare Studies serves as a think tank whose purpose includes developing new war-fighting concepts, linking strategic matters with technological developments, and fostering college curriculum development.[81]

NWC's Web site contains a variety of information resources on naval doctrine. One example is the Current Strategy Forum, which is an annual exchange of views by civilian and military leaders on major national and international strategic issues and the roles maritime forces can play in addressing these matters. The Naval War College Press (http://www.nwc.navy.mil/press/) publishes valuable resources in this area, including the scholarly journal, *Naval War College Review* (http://purl.access.gpo.gov/GPO/LPS17060 (2004–present) and http://purl.access.gpo.gov/GPO/LPS95072 (1996–2004)), and the Newport Papers monographic series, whose representative titles include *The Doctrine Reader: The Navies of the United States, Great Britain, France, Italy, and Spain* (1995), *The Evolution of the U.S. Navy's Maritime Strategy, 1977–1986* (2004), *Naval Power in the 21st Century: A Naval War College Review Reader* (2005), and *Shaping the Security Environment* (2007).

NWC's China Maritime Studies Institute (http://www.nwc.navy.mil/cnws/cmsi/) seeks to understand and analyze China's increasing international maritime importance, and its Web site provides citations and links to some publications on Chinese naval trends and developments.

Naval Postgraduate School

The Naval Postgraduate School (NPS) (http://www.nps.edu/) seeks to provide pertinent and unique advanced education and research programs to enhance the combat effectiveness of U.S. and allied armed forces, while also enhancing U.S. national security.[82] There are a number of NPS research institutes that produce publications dealing with military or naval doctrine and strategy. Examples of these institutes include the Center for Civil-Military Relations (http://www.ccmr.org/), the Center for Contemporary Conflict (http://www.ccc.nps.navy.mil), Center for

Homeland Defense and Security (http://www.chds.us/), Center for Stabilization and Reconstruction Studies (http://www.csrs-nps.org/), Center for Survivability and Lethality (http://www.nps.edu/academics/GSEAS/MAE/CSL/), Center for Terrorism and Irregular Warfare (http://www.nps.edu/academics/centers/CTIW/), and Program for Culture and Conflict Studies (http://www.nps.edu/Programs/CCS/). Examples of publications produced by these organizations include the journals *Culture and Conflict Review* (November 2007–present) and *Strategic Insights* (March 2002-present) and reports or student theses such as "The Future of Armed Resistance: Cyberterror? Mass Destruction?" (2000), "An Alternate Military Strategy for the War on Terrorism" (2004), "Falling out of Formation: A Look at the Navy's Search for a New Maritime Strategy" (2007), and "North Korea's Juche Ideology and the German Reunification Experience" (2008).

NPS's Homeland Security Digital Library (HSDL) (http://www.hsdl.org/) is also a good resource for documents on homeland security, including those covering naval or maritime doctrine. *Naval Cooperation after Korean Unification* (1995), *In Search of an Operational Doctrine for Maritime Counterterrorism* (2003), and *The Growth of the Chinese People's Liberation Army Navy: Impacts and Implications of Regional Naval Expansion* (2007) are examples of relevant HSDL naval doctrine resources.

Center for Naval Analyses

The Center for Naval Analyses (CNA) (http://www.cna.org/) is an Alexandria, VA-based nonprofit research organization providing empirical professional analysis of various national security, international affairs, and assorted public policy issues.[83] Examples of pertinent naval doctrine and strategic products prepared by CNA include *Forward... From the Start: The U.S. Navy and Homeland Defense, 1775–2003* (2003), *China's Revolution in Doctrinal Affairs: Emerging Trends in the Doctrinal Art of the Chinese People's Liberation Army* (2005), *The Future of U.S. Deterrence: Constructing Effective Strategies to Deter States and Non-State Actors* (2007), *U.S. Navy Capstone Strategies, Visions, & Concepts (1970–2008) With Insights for the U.S. Navy of 2009 & Beyond* (2008), and *Report on the Gulf Naval Commanders Conference* (2008).

Johns Hopkins University Applied Physics Laboratory: Rethinking Maritime Strategy

The Johns Hopkins University Applied Physics Laboratory Maritime Strategy program (http://www.jhuapl.edu/maritimestrategy/) seeks to analyze and promote discussion of future elements that should be included in U.S. Navy maritime strategic development. Topics discussed as part of this initiative include collecting inputs and analyzing the strategic maritime environment; developing maritime strategies; testing, examining, and refining alternatives; and synthesizing and reporting development principles to sustain this strategy's value and legitimacy.[84]

Comments on proposed maritime strategy are posted by individuals such as former Chief of Naval Operations Admiral Mike Mullen (2006), and these comments include observations by other interested individuals who wish to foster additional discussion of these subjects. Categories of discussion topics and comments on this Web site include views on current Navy strategic documents; protecting, monitoring, and controlling the Exclusive Economic Zone; piracy; smuggling of people, weapons, and drugs; sea-lane security; port and harbor security; and U.S. maritime industrial base security and capability.

An additional resource for analysis of U.S. naval doctrine and strategy is the periodical, *Proceedings of the U.S. Naval Institute,* which features articles on a variety of naval subjects, including strategic and doctrinal matters. Publications from 1996–present are accessible at http://www.usni.org/magazines/proceedings/.

Although some U.S. military doctrine documents are inaccessible for national security reasons, the vast majority of current U.S. military doctrine and national security strategy documents are publicly available. This enables interested readers to actually read these documents and understand the rationales that military and civilian document writers present to explain why U.S. military forces seek to conduct military operations in particular ways. This transparency and multifaceted access makes the U.S. military the world's leader in providing information about its military doctrine to individuals interested in studying and analyzing this critically important topic.

Notes

1. Jay M. Shafritz, Todd J. A. Shafritz, and David B. Robertson, *The Facts on File Dictionary of Military Science* (New York: Facts on File, 1989), 246. Other historical descriptions of joint military doctrine as applied to the U.S. military include Roger D. Launius, "Military Unification's Precursor: The Air Force and Navy Strategic Airlift Merger of 1948," *Air Power History* 39, no. 1 (1992): 22–33; David Jablonsky, "Eisenhower and the Origins of Unified Command," *Joint Force Quarterly* 23 (1999–2000): 24–31; James R. Locher, *Victory on the Potomac: The Goldwater-Nichols Act Unifies the Pentagon* (College Station: Texas A&M University Press, 2002); James A. Kitfield, "A Better Way to Run a War," *Air Force Magazine* 89, no. 10 (2006): 36–40; and Michael C. Veneri, "The U.S. Military's Implementation of the Joint Duty Promotion Requirement," *Armed Forces and Society* 34, no. 3 (2008): 413–432.

2. For a description of the Superintendent of Documents (SuDoc) classification system, see U.S. Government Printing Office, Federal Depository Library Program, "An Explanation of the Superintendent of Documents Classification System," (Washington, DC: Government Printing Office, 2004), http://www.access.gpo.gov/su_docs/fdlp/pubs/explain.html (accessed October 20, 2008); and Donna Burton, ed., *Guide to U.S. Government Publications* (Detroit: Gale Group, 2008).

3. There are no scholarly articles in library science literature examining the role of national security strategy documents as research tools. A partially related article on presidential national security directives is Catherine M. Dwyer, "The U.S. Presidency and National Security Directives: An Overview," *Journal of Government Information* 29, no. 6 (2002): 410–419. Bert Chapman, *Researching National Security and Intelligence Policy*

(Washington, DC: CQ Press, 2004) is an example of work on conducting library research with national security documentation.

4. President of the United States, *National Security Strategy of the United States* (Washington, DC: White House, 1987), 1.

5. Ibid., 4, 13, 16–17, 21–23, 26–31, and 35–40.

6. President of the United States, *A National Security Strategy of Engagement and Enlargement* (Washington, DC: White House, 1995), 10–11, 13, and 20–21.

7. President of the United States, *A National Security Strategy for a Global Age* (Washington, DC: White House, 2000), 9, 31–33, and 36.

8. President of the United States, *National Security Strategy of the United States* (Washington, DC: White House, 2002), 1–2.

9. Ibid., 6.

10. Literature on the Bush Administration's preemptive doctrine includes Chris J. Dolan, *In War We Trust: The Bush Doctrine and the Pursuit of Just War* (Burlington, VT: Ashgate Publications, 2005); Gary Rosen, ed., *The Right War: The Conservative Debate on Iraq* (New York: Cambridge University Press, 2005); James Kitfield, *War and Destiny: How the Bush Revolution in Foreign and Military Affairs Redefined American Power* (Washington, DC: Potomac Books, 2005); William W. Keller and Gordon R. Mitchell, eds., *Hitting First: Preventive Force in U.S. Security Strategy* (Pittsburgh: University of Pittsburgh Press, 2006); Lyle Goldstein, *Preventive Attack and Weapons of Mass Destruction: A Comparative Historical Analysis* (Stanford, CA: Stanford University Press, 2006); and Robert G. Kaufman, *In Defense of the Bush Doctrine* (Lexington: University Press of Kentucky, 2007).

11. President of the United States, *National Security Strategy of the United States* (Washington, DC: The White House, 2006), 12.

12. See United States Department of Defense, *Quadrennial Defense Review Report* (Washington, DC: Department of Defense, 2001), 71 and 10 *USC 118*.

13. United States Department of Defense, *Quadrennial Defense Review Report* (Washington, DC: Department of Defense, 2006), vi.

14. Ibid., vii.

15. *United States Government Manual, 2008–2009* (Washington, DC: Government Printing Office, 2008), 156.

16. U.S. Joint Chiefs of Staff, *National Military Strategy of the United States* (Washington, DC: Department of Defense, 1992), 1.

17. U.S. Joint Chiefs of Staff, *National Military Strategy of the United States of America: A Strategy for Today; A Vision for Tomorrow* (Washington, DC: Department of Defense, 2004), 1–27.

18. U.S. Department of Defense, *The National Defense Strategy of the United States of America* (Washington, DC: Department of Defense, 2005), iv.

19. United States, Joint Chiefs of Staff, *National Military Strategic Plan for the War on Terrorism* (Washington, DC: 2006), 13.

20. Ibid., 6–8.

21. United States, Joint Chiefs of Staff, *National Military Strategy to Combat Weapons of Mass Destruction* (Washington, DC: 2006), 7.

22. Ibid., 13.

23. Ibid., 13–17.

24. Alice C. Cole et al., eds., *The Department of Defense: Documents on Establishment and Organization, 1944–1978* (Washington, DC: Office of the Secretary of Defense, Historical Office, 1978), 175.

25. David Jablonsky, "Eisenhower and the Origins of Unified Command," *Joint Force Quarterly* 23 (1999–2000): 30–31.

26. The most authoritative review of Goldwater-Nichols background is James R. Locher III, *Victory on the Potomac: The Goldwater-Nichols Act Unifies the Pentagon* (College Station: Texas A&M University Press, 2002). The National Defense University Library, http://www.ndu.edu/library/goldnich/goldnich.html, features the full text of congressional committee hearings on this legislation from 1981–1988. Assessments of Goldwater-Nichols effectiveness include Christopher Bourne, "Unintended Consequences of the Goldwater-Nichols Act," *Joint Force Quarterly* 18 (1998): 99–108; Peter J. Roman and David W. Tarr, "The Joint Chiefs of Staff: From Service Parochialism to Jointness," *Political Science Quarterly* 113, no. 1 (1998): 91–111; Dennis J. Quinn, ed., *The Goldwater-Nichols DOD Reorganization Act: A Ten-Year Retrospective* (Washington, DC: National Defense University Press, 1999); Chuck Harrison, "How Joint Are We and Can We Be Better?," *Joint Force Quarterly* 38 (2005): 14–19; and James R. Locher III, "Has It Worked?: The Goldwater-Nichols Act Reorganisation Act," *Air Power Journal* 1, no. 2 (2006): 155–179.

27. National Defense University Library, "Goldwater Nichols Department of Defense Reorganization Act of 1986," (n.d.), http://www.ndu.edu/library/goldnich/goldnich.html (accessed October 8, 2008).

28. For a chart of these publications, see Joint Electronic Library, Joint Doctrine Branch, "Publications Hierarchy Chart," http://www.dtic.mil/doctrine/publicationshierarchychart.htm (accessed October 8, 2008).

29. U.S. Joint Chiefs of Staff, *JP 1 Joint Doctrine for the Armed Forces of the United States* (Washington, DC: JCS, 2007), v–viii.

30. Ibid., I-2.

31. Ibid., II-1.

32. U.S. Joint Chiefs of Staff, *JP 3–18 Joint Forcible Entry Operations* (Washington, DC: JCS 2008), v–vi.

33. Ibid., vii–viii.

34. Examples of this burgeoning field of scholarly analysis include Robert Frank Futrell, *Ideas, Concepts, Doctrine: Basic Thinking in the United States Air Force* (Maxwell Air Force Base, AL: Air University Press, 1989); Charles M. Westerhoff, comp., *Military Airpower: The CADRE Digest of Air Power Opinions and Thought* (Maxwell Air Force Base, AL: Air University Press, 1990); James A. Mowbray, "Air Force Doctrine Problems: 1926-Present," *Airpower Journal* 9, no. 4 (1995): 21–41; Dennis M. Drew, "U.S. Airpower Theory and the Insurgent Challenge: A Short Journey to Confusion," *Journal of Military History* 62, no. 4 (1998): 809–832; Philip S. Meilinger, *Airmen and Air Theory: A Review of the Sources* (Maxwell Air Force Base, AL: Air University Press, 2001); Clayton K.S. Chun, *Aerospace Power in the 21st Century: A Basic Primer* (Colorado Springs and Maxwell Air Force Base, AL: U.S. Air Force Academy and Air University Press, 2001); John T. Correll, "Basic Beliefs: Recent Decades Have Brought Some Major Changes in Air Force Doctrine," *Air Force Magazine* 87, no. 6 (2004): 42–47; Irving B. Holley Jr., *Technology and Military Doctrine: Essays on a Changing Relationship* (Maxwell Air Force Base, AL: Air University Press, 2004); and Bruce R. Pirnie et al., *Beyond Close Air Support: Forging a New Air-Ground Partnership* (Santa Monica, CA: Rand Corporation, 2005).

35. U.S. Air Force, *Air Force Basic Doctrine AFDD 1* (Washington, DC: U.S. Air Force, 2003), iii–v.

36. Ibid., 4–7, 9.

37. Ibid., 15.

38. Ibid., 80.

39. U.S. Air Force, *Nuclear Options AFDD 2–1.5* (Washington, DC: U.S. Air Force, 1998), iii–iv.

40. Ibid., v–vi.

41. Ibid., 1–2.

42. Ibid., 8–9.

43. Air University, Lemay Center for Doctrine Development and Education (n.d.), 1, http://www.cadre.maxwell.af.mil/about.asp (accessed October 9, 2008).

44. Rand Corporation, Project Air Force, "Fiscal Year 2008 Research Agenda," http://www.rand.org/paf/agenda/stratdoc.html (accessed October 9, 2008).

45. A representative sampling of this proliferating literature on U.S. Army doctrine includes Russell F. Weigley, *The American Way of War: A History of United States Military Strategy and Policy* (Bloomington: Indiana University Press, 1977); John P. Rose, *The Evolution of U.S. Army Nuclear Doctrine, 1945–1980* (Boulder, CO: Westview Press, 1980); Dennis Stewart Diggers, "The United States Army's Long March from Saigon to Baghdad: The Development of War Fighting Doctrine in the Post–Vietnam Era" (PhD diss., Syracuse University, 1996); Walter Edward Kretchik, "Peering Through the Mist: Doctrine as a Guide for U.S. Army Operations" (PhD diss., University of Kansas, 2001); Antulio J. Echevarria II, *Clausewitz's Center of Gravity: Changing Our Warfighting Doctrine—Again!* (Carlisle, PA: Strategic Studies Institute, U.S. Army War College, 2002); James F. Gebhardt, *The Road to Abu Ghraib: US Army Detainee Doctrine and Experience* (Fort Leavenworth, KS: Combat Studies Institute Press, 2005); Andrew James Birtle, *U.S. Army Counterinsurgency and Contingency Operations Doctrine, 1942–1976* (Washington, DC: U.S. Army Center of Military History, 2006); Colin S. Gray, *The Implications of Preemptive and Preventive War Doctrines* (Carlisle, PA: Strategic Studies Institute, U.S. Army War College, 2007); and Rudolph M. Janiczek, *A Concept at the Crossroads: Rethinking the Center of Gravity* (Carlisle, PA; Strategic Studies Institute, U.S. Army War College, 2007).

46. Search conducted by author September 25, 2008.

47. U.S. Army, *FM 1 The Army: Our Army at War Relevant and Ready Today and Tomorrow* (Washington, DC: U.S. Army, 2005), 4–2.

48. Ibid.

49. U.S. Army, *FM 3–24 Counterinsurgency* (Washington, DC: U.S. Army, 2006), i–v.

50. Ibid., 7–5 to 7–6.

51. Ibid., 7–6.

52. See National Archives and Records Administration, *United States Government Manual, 2008–2009* (Washington, DC: Government Printing Office, 2008), 172; Norma Vishneski, ed., *TRADOC, U.S. Army Training and Doctrine Command: Preparing for the Future* (Fort Monroe, VA: The Command, 1981), introduction and chapter 3; and Anne Chapman, Benjamin King, Carol Lilly, and John Romjue, *Transforming the Army: TRADOC's First Thirty Years, 1973–2003* (Fort Monroe, VA: United States Army Training and Doctrine Command, Military History Office, 2003).

53. See "Army Capabilities Integration Center," (n.d.), http://www.arcic.army.mil; "CAC Overview," (2008), http://usacac.army.mil/CAC2/overview.asp; and "United States Army Combined Arms Support Command," (2008), http://www.cascom.lee.army.mil/default.asp (accessed October 14, 2008).

54. U.S. Army War College, Strategic Studies Institute, "About the Strategic Studies Institute," (n.d.), http://www.strategicstudiesinstitute.army.mil/about/strategic-studies-institute.cfm (accessed October 14, 2008).

55. Combat Studies Institute, "About CSI," (2008), http://usacac.army.mil/cac2/csi/aboutCSI.asp (accessed October 14, 2008).

56. Association of the United States Army, "What is AUSA?," (2008), http://www.ausa.org/about/what/Pages/default.aspx (accessed October 15, 2008).

57. Association of the United States Army, "Institute of Land Warfare," (2008), http://www.ausa.org/about/ilw/Pages/default.aspx (accessed October 15, 2008).

58. Rand Corporation, "About Arroyo Center," (2008), http://www.rand.org/ard/about.html (accessed October 15, 2008).

59. Examples of these appraisals of Marine Corps doctrine include U.S. Marine Corps, *Small Wars Manual* (Washington, DC: U.S. Marine Corps, 1940); Allan S. Millett, *Semper Fidelis: The History of the United States Marine Corps* (New York: Macmillan, 1980); Robert S. Trout, "Dysfunctional Doctrine: The Marine Corps and FMFM1 Warfighting," *Marine Corps Gazette* 77, no. 10 (1993): 33–35; Stephen L. Goertzen, "The Feasibility of the Over-the-Horizon Amphibious Assault for U.S. Navy and Marine Corps Forces" (master's thesis, U.S. Army Command and General Staff College, 1993); Garrett J. Sullivan, "The Genesis of Amphibious Warfare Doctrine," *Military Review* 75, no. 3 (1995): 95–97; United States Marine Corps, *Marine Corps Doctrine* (Washington, DC: Headquarters, United States Marine Corps, 1997); David Keithly and Paul Melshen, "Past as Prologue: USMC Small Wars Doctrine," *Small Wars and Insurgencies* 8, no. 2 (1997): 87–108; Keith B. Bickel, *Mars Learning: The Marine Corps Development of Small Wars Doctrine, 1915–1940* (Boulder, CO: Westview Press, 2001); John C. Madsen, *Reorganization of the Marine Air Command and Control System to Meet 21st Century Doctrine and Technology* (Monterey, CA: Naval Postgraduate School, 2001); and Terry Terriff, "'Innovate or Die': Organizational Culture and the Origins of Maneuver Warfare in the United States Marine Corps," *Journal of Strategic Studies* 29, no. 3 (2006): 475–503.

60. USMC Artillery Detachment, Fort Sill, OK, "Marine Corps Publications Lead Series," (2008), http://sill-www.army.mil/USMC/Pubs/ (accessed October 15, 2008).

61. U.S. Marine Corps, *MCDP 3 Expeditionary Operations* (Washington, DC: U.S. Marine Corps, 1998), 35–36.

62. Ibid., 36.

63. U.S. Marine Corps, *MCWP 2–14 Counterintelligence* (Washington, DC: U.S. Marine Corps, 2000), 2–1.

64. Ibid.

65. U.S. Marine Corps Combat Development Command, "Homepage," (2008), https://www.mccdc.usmc.mi/ (accessed October 16, 2008).

66. U.S. Marine Corps Combat Development Command, Operations Analysis Division, "Who Are We?," (n.d.), https://www.mccdc.usmc.mil/OperationsAnalysis/default.asp (accessed October 16, 2008).

67. "The Marine Corps Small Wars Center of Excellence," *Marines Corps Gazette* (July 2005): 37–38.

68. U.S. Marine Corps, Center for Occupational Cultural Learning, "Mission," (2008), http://www.tecom.usmc.mil/caocl/ (accessed October 16, 2008).

69. Marine Corps University, "MCU Vision Statement," (n.d.), http://www.mcu.usmc.mil/mcu/mission_vision/mission_vision.htm (accessed October 16, 2008).

70. Marine Corps University Press, "Marine Corps University Journal," (n.d.), http://www.tecom.usmc.mil/mcupress/journal.htm (accessed April 29, 2009).

71. Demonstrations of this literature on U.S. naval doctrinal development include Ronald Spector, *Professors of War: The Naval War College and the Development of the Naval*

Profession (Newport, RI: Naval War College Press, 1977); George W. Baer, *One Hundred Years of Sea Power: The U.S. Navy, 1890–1990* (Stanford, CA: Stanford University Press, 1994); James John Tritten, *Development Issues for Multinational Navy Doctrine* (Norfolk, VA: Naval Doctrine Command, 1995); Sam J. Tangredi, ed., *Globalization and Maritime Power* (Washington, DC: National Defense University Press, 2002); R. Blake Dunnavent, *Brown Water Warfare: The U.S. Navy in Riverine Warfare and the Emergence of a Tactical Doctrine, 1775–1970* (Gainesville: University Press of Florida, 2002); John B. Hattendorf, *The Evolution of the U.S. Navy's Maritime Strategy, 1977–1986* (Newport, RI: Naval War College, Center for Naval Warfare Studies, 2004); Peter Dombrowski, ed., *Naval Power in the 21st Century: A Naval War College Review Reader* (Newport, RI: Naval War College Press, 2005); John B. Hattendorf, ed., *U.S. Naval Strategy in the 1990s: Selected Documents* (Newport, RI: Naval War College Press, 2006); W. J. Holland, "Challenges for the New Maritime Strategy," *U.S. Naval Institute Proceedings* 133, no. 4 (2007): 14–18; and Andrew Lambert, "Strategy," in *The Oxford Encyclopedia of Maritime History*, ed. John B. Hattendorf (New York: Oxford University Press, 2007), 3: 56–57.

72. U.S. Navy, *Naval Doctrine Publication 1 Naval Warfare* (Washington, DC: Department of the Navy, Office of the Chief of Naval Operations, 1994), 1.

73. Ibid., 3.

74. Ibid., 8.

75. Ibid., 13.

76. U.S. Navy, *Naval Doctrine Publication 6 Naval Command and Control* (Washington, DC: Department of the Navy, Office of the Chief of Naval Operations, 1995), 18–19.

77. Ibid., 32.

78. U.S. Naval Warfare Development Command, "Homepage," (2008), http://www.nwdc.navy.mil/ (accessed October 17, 2008).

79. U.S. Naval Warfare Development Command, "Sea Power 21," (n.d.), http://www.nwdc.navy.mil/content/conops/Seapower21.aspx (accessed October 17, 2008). The full text of *Sea Power 21* can be found at http://www.navy.mil/navydata/cno/Proceedings.html (accessed October 17, 2008).

80. U.S. Navy, "Responsibilities of the Chief of Naval Operations," (n.d.), http://www.navy.mil/navydata/navy_legacy_hr.asp?id=239 (accessed October 17, 2008).

81. U.S. Naval War College, "Overview: Greetings from the Naval War College," (n.d.), http://www.nwc.navy.mil/about/ (accessed October 21, 2008). For an overview of NWC's origins, see Ronald Spector, *Professors*.

82. U.S. Naval Postgraduate School, "Naval Postgraduate School Mission Statement," (n.d.), http://www.nps.edu/Aboutnps/ (accessed October 21, 2008).

83. Center for Naval Analyses, "CNA: About Us," (n.d.), http://www.cna.org/about/ (accessed October 21, 2008).

84. Johns Hopkins University Applied Physics Laboratory, "About Rethinking Maritime Strategy," (2006–2007), http://www.jhuapl.edu/maritimestrategy/about.htm (accessed October 1, 2008).

CHAPTER 3

Foreign Government Military Doctrine Resources

This chapter will examine and describe the military doctrine resources produced by foreign governments and militaries. Different historical, political, and military factors are involved in the production of these information resources. Some are statements of overall national military policy and strategy while others are expressions of how individual or joint armed services conduct military operations in certain areas, such as how armies conduct armored operations. As a general rule, such public doctrinal or policy statements are more likely to be produced by democratic governments than nondemocratic regimes.

Both civilian and military policymakers may be involved in developing these doctrinal statements, similarly to how U.S. civilian policymakers may involve executive or legislative branch officials.[1] This chapter will look at recent military doctrinal documents produced by a representative sampling of countries from around the world. Emphasis will be placed on documents that are Internet accessible and available in English. Excellent gateways to foreign military doctrine documents are provided by the Online Public Access Catalog (OPAC) of the U.S. Army War College Library, which is accessible at http://www.carlisle.army.mil/library/, and the National Defense University's Military Education and Research Library Network's (MERLN) White Papers on Defense (http://merln.ndu.edu/whitepapers.html). There are two ways of searching the Army War College OPAC for foreign military doctrine documents. The first is to do a Library of Congress Subject Headings (LCSH) search under the phrase, "military doctrine," which will produce a results list that will contain records for military doctrine documents from many countries. If you are interested in Brazilian military doctrine, one of the results you get will say "military doctrine—Brazil." This entry will connect you to the catalog record for a 2005 Brazilian Ministry of Defense national policy document and provide a link to an online version of this document.

Another way of searching the documents in the U.S. Army War College OPAC is through a title search under the phrase, "White Papers on Defense," which will produce more than one hundred results. Besides cataloging the titles of books, it is also possible to use library OPACs to search for monographic series by particular publishers to see all the works produced by that publisher on a particular topic that are available in that library. The Army War College Library has cataloged many of these works under the series, "White Papers on Defense," to facilitate user access to these publications.

Online versions of some of these publications may be available through the Web sites of the issuing national defense ministry, armed service branches, or civilian agency, and they are, in some cases, available on a link provided through the National Defense University Library at Fort McNair in Washington, DC. This chapter will now look at these documents, describe their contents, the organizational entities involved in producing these resources, and provide the web Uniform Resource Locators (URLs) where they can be found. In researching and writing this work, the author has made a good faith effort to provide access to the most recent versions of these documents available.

The following section of this chapter will examine military doctrine and strategy documents produced by countries other than the United States and describe the multiple political and military factors responsible for their creation.

Australia

Australian national military strategic and military doctrine documents are produced by a number of entities, including the Department of Defence, its armed services, including branches of those services such as the Royal Australian Air Force's Airpower Development Centre, the Army's Land Warfare Studies Centre, and the Royal Australian Navy's Seapower Centre. These documents will reflect experience gleaned from Australia's remarkable history of military operations,[2] along with ongoing operations in areas as diverse as Afghanistan and East Timor, and review future security threats that may require committing Australian military forces in order to defeat these threats. Australian military doctrine documents will reflect joint national military perspectives and the perspectives of individual branches of its armed services.

Australia's Department of Defence is the first place to begin our search for Australian military strategic and doctrine documents. Their Web site (http://www.defence.gov.au/) is the place to begin, and the Reports and Publications section of this Web site features a cornucopia of documents. One document to initially consult is the 2000 Defence White Paper, *Defence 2000: Our Future Defence Force.* Prepared by the Conservative Coalition Government of Prime Minister John Howard, who was in power from 1996–2007, this document stresses that it was compiled by extensive governmental, military, and public consultation and that its goal is to explain Australian defense and strategic policies to Australia's allies and neighbors in the hope of promoting greater understanding of Australian security interests and preventing misunderstandings.[3]

Other noteworthy statements of Australian national military strategy and doctrine that emphasize joint service collaboration and analysis of that strategy and doctrine on the Defence Department Web site include documents such as *Defence Annual Reports* (1997/1998–present), articles from the *Australian Defence Force Journal* (1997–present), *Australian Approach to Warfare* (2002), *Force 2020* (2002), *Defence Update: Australia's National Security* (2007), *Joint Operations for the 21st Century* (2007), and *Network Centric Warfare Map* (2007).

The 2007 election victory of the Australian Labour Party and Prime Minister Kevin Rudd brought a new government to power, and it is currently in the process of drafting a new defense White Paper to stress its national security policies and priorities. A section of the departmental Web site (http://www.defence.gov.au/whitepaper/) created in 2008 announces public meetings at various locations to solicit feedback on what should be in this forthcoming document. An interactive feature is provided to give interested individuals the opportunity to submit their suggestions and recommendations for the white paper.

Each of Australia's individual armed services also produces resources on the military doctrine of their respective branches, including the text of doctrine documents as well as discussion and analysis of these resources. The Royal Australian Air Force's (RAAF) Air Power Development Centre (http://www.raaf.gov.au/airpower/) has a number of useful resources. These include the four keystone documents of Australian airpower doctrine:

- *AAP 1000D Air Power Manual* (2007), which stresses the role of air and space power in Australian national security;
- *AAP 1000F Future Air and Space Operating Concept* (2007), which emphasizes the roles played by command and control, information superiority and support, and force application and sustainment in national aerospace operations;
- *AAP 1000H Australian Experience of Air Power* (2007), which reviews the historical development of Australian military air power; and
- *AAP 1003 Operations Law for RAAF Commanders* (2004), which covers topics such as the legal division between airspace and oceans, aerial targeting law, adhering to and enforcing the law of armed conflict, and the legal role of deception in armed conflict.

The Airpower Development Centre Web site also features papers such as *Putting Space into RAAF Aerospace Power Doctrine* (2003), working papers such as *Operational Level Doctrine: Planning an Air Campaign* (1993), and the text of selected other publications.

The Australian Army's Land Warfare Studies Centre (http://www.defence.gov.au/lwsc/) serves as the Army's think tank, providing a variety of resources on Australian Army doctrine. These include articles from the *Australian Army Journal* (June 2003–present), *Senior Officer's Professional Digest*, which summarizes articles from a variety of global professional military journals (2002–present), Study Papers, such as *Forward from the Past: The Development of Australian Army Doctrine 1972–Present* (1999), and Working Papers, such as *Revisiting Counterinsurgency: A Manoeuverist Approach Response to the 'War on Terror' for the Australian Army* (2006).

The keystone Australian army doctrinal publication, *Land Warfare: Fundamentals of Land Warfare LWD 1*, can be found on the Australian Army Web site at (http://www.defence.gov.au/army/LWD1/), and its contents include chapters covering topics such as influences on modern land warfare, military strategy, conducting land warfare, and generating land warfare capability. An excerpt from the first chapter of this document describes asymmetric warfare as follows:

> Asymmetric warfare describes military actions against an adversary to which he may have no effective response and which pit strength against weakness, sometimes in a non-traditional and unconventional manner. In terms of the application of land power, it is important to draw a distinction between asymmetric warfare as employed by the militaries of modern liberal democracies and asymmetric warfare as employed by their real and potential opponents. In the context of military operations by modern liberal democratic states, the aim of asymmetry is to achieve disproportionate effects and to afford an enemy no effective counter to the forces used against him.[4]

The Royal Australian Navy's Seapower Centre Australia (http://www.navy.gov.au/spc/) serves as the agency responsible for developing Australian maritime power and Australian naval doctrine and incorporating that doctrine into Australian joint military strategy.[5] Publications here include the keystone information resource, *Australian Maritime Doctrine RAN Doctrine 1* (2000), whose contents include the political, economic, and social factors affecting Australia's maritime environment relationships; the origins of maritime strategic thought and how it affects current and future maritime strategic concepts; the operational relationship between air, land, and sea forces; and characteristics of maritime organization and campaigning.

Additional documentary resources on this Web site include *The Navy Contribution to Australian Maritime Operations: RAN Doctrine 2* (2005), the newsletter *Semaphore,* which describes historic and current Australian naval operations (2002–present), *Working Papers* from 1999–present, which include *An Effects-Based Anti-Submarine Warfare Strategy* (2006), and the *Papers in Australian Maritime Affairs* series (1996–present) including *Freedom of Navigation in the Indo-Pacific Region* (2008).

Other Australian sources evaluating Australian military doctrine and national military strategy include the Army's Center for Army Lessons (http://www.defence.gov.au/army/cal/), publications produced at the Australian Defence College (http://www.defence.gov.au/adc/). These include *Occasional Series* publications such as *City Without Joy: Urban Military Operations in the 21st Century* (2007) and the *Monograph Series,* which includes *The Personnel Dimension of ADF Capability: Future Vulnerability or Strength?* (2004), and the Australian Strategic Policy Institute (http://www.aspi.org.au/), whose pertinent publications include *ADF Capability Review: Royal Australian Air Force* (2008) and *Asian Military Trends and Their Implications for Australia* (2008), and Australian National University's Strategic and Defence Studies Centre (http://rspas.anu.edu.au/sdsc/).

All of these resources demonstrate that Australia is a model of transparency in providing information about national military strategy and doctrine and the doctrine of its individual armed services.

Brazil

Brazilian military doctrine and policy have been influenced by that country's complicated history of civil-military relationships, which have included extended periods of military rule.[6] Brazil's most recent national military policy document published in English is its 2005 *National Defense Policy,* available through National Defense University's Library at http://merln.ndu.edu/whitepapers/Brazil_English2995.doc. This document stresses international security environment characteristics, such as the development of globalization, the increasing importance of environmental issues, continuing advancements in science and technology, including satellites and electronic sensing devices, the increasing importance of non-governmental actors in international security, and the increasing threats to global security posed by transnational crime and terrorism.[7]

This document goes on to stress the importance of the South American subcontinent as the regional security environment where Brazil is most likely to intervene, and also stresses that national policymakers envision Brazilian strategic interests as encompassing the South Atlantic border and adjacent African countries. Further, this document emphasizes that Brazil seeks to reduce the possibility of conflicts in this region through its involvement in organizations such as Mercosur and the South American Community of Nations; that the Brazilian Amazon's mineral and biodiversity wealth potential need better defenses and demarcation against transnational crime; that access to oceanic resources is becoming increasingly important to national economic development and national security; that Brazil seeks to defend an international order based on democracy, multilateralism, cooperation, and peaceful dispute resolution; and that it seeks to enhance its defense capabilities with ongoing involvement from its government, business, and academic sectors.[8]

Additional information and discussion of Brazilian military doctrine can be found in resources produced by National Defense University's Center for Hemispheric Defense Studies (http://www.ndu.edu/chds/), including the e-journal *Security and Defense Studies Review,* which features articles in Spanish and Portuguese, and from the Portuguese language resources of Brazil's Defense Ministry (http://www.defesa.gov.br/). Such resources provide additional insight into the military policy thinking of South America's most powerful country. It is also possible that future Brazilian writings on this subject will focus on whether the policies of leaders such as Bolivia's Evo Morales and Venezuela's Hugo Chavez will be detrimental to Brazilian national security interests.

Canada

Early 21st-century Canadian military doctrine has been influenced by that country's historically close ties to France, Great Britain, and the United States. This was particularly reflected in Canadian participation in two world wars and in the Korean War.[9] Since these conflicts, Canadian military policy and doctrine has placed great emphasis on serving in United Nations international peacekeeping operations; however, Canada's ongoing involvement in combat operations in Afghanistan, which emphasize counterinsurgency activities, and the desire of Prime Minister Stephen Harper's government to increase the size of Canada's military may herald a more robust posture by the Canadian military in years to come.[10]

Canadian military doctrine documents may be found in many areas, with the Department of National Defence (DND) Web site (http://www.dnd.ca/) being an important place to start. The Defence Policy Archives section of DND's Web site is an excellent place to begin because it contains the full text of eight Canadian national military strategy documents from the 1960s to the present. Examples of these documents include *White Paper on Defence* (1964), *Challenge and Commitment* (1987), *Defence Policy White Paper* (1994), and *Canada First Defence Strategy* (2008). This last document was produced in June 2008 by the Harper government, and Defence Minister Peter Mackay was also responsible for its preparation.

Canada First Defence Strategy reflects the government's desire and commitment to gradually increase defense spending and the size of Canadian forces. Capabilities desired from this enhanced fiscal support include the abilities to conduct daily domestic and continental operations in the Arctic and through North American Aerospace Defense Command (NORAD); respond to a major terrorist attack; lead and/or conduct a major international operation for an extended period; support Canadian civilian authorities if a natural disaster occurs; and deploy forces to respond to global security crises for shorter periods.[11]

Numerous additional resources provide access to Canadian military doctrine documents and analyses of this doctrine. A place to start is the Canadian Forces Joint Doctrine Branch (http://www.cfd-cdf.forceds.gc.ca/sites/page-eng.asp?page=3047), which features the text of many documents emphasizing how Canadian military forces conduct operations by themselves and with allied countries. Examples of these publications include *Canadian Forces Joint Doctrine for Mobilization (FP-020)*(2002) and *Non-Combatant Evacuation Operations (FP-050)* (2003). The *Canadian Military Journal* (http://www.journal.forces.gc.ca/) covers features articles on Canadian military policy from 2000–present and also provides analysis of these issues.

Canadian Air Force doctrinal resources can be found through the Canadian Forces Aerospace Warfare Centre (http://www.airforce.forces.gc.ca/cfawc/). Pertinent materials available here include keystone documents such as *Canadian Forces Aerospace Doctrine* (2006) and supplemental analyses that include *Canadian Air Force Leadership and Command: The Human Dimension of Expeditionary Air Force Operations* (2007) and *Command and Control of Canadian Aerospace Forces: Conceptual Foundations* (2008).

Applicable Canadian Army resources may be found through the Army's Web site (http://www.army.forces.gc.ca/). This site includes articles from *Canadian Army Journal* (http://www.army.forces.gc.ca/CAJ/), whose coverage dates from 1998–present. Canadian Navy doctrinal information can be found within sections of its Web site (http://www.navy.dnd.ca/).

Further analyses of Canadian military doctrine are provided by Canada's professional military educational institutions. These analyses include documents produced at the Canadian Defence Academy (http://www.cda-acd.forces.gc.ca/CLFI/engraph/research/research_e.asp) and the Canadian Forces College, which provides papers from 1998–present at http://wps.cfc.forces.gc.ca/en/cfpapers/.

Numerous Canadian academic institutions have centers of expertise that analyze current and future defense issues, and some of this research is published. Examples of such publications are produced by the University of Calgary's Centre for Military and Strategic Studies (http://www.cmss.ucalgary.ca/), The University of New Brunswick's Gregg Centre for the Study of War and Society (http://www.unb.ca/greggcentre/), and the Queen's University School of Policy Studies Defence Management Studies Program (http://www.queensu.ca/sps/defence_management/).

Consequently, Canadian government and civilian organizations are a rich source of military doctrine documentation and analysis.

China

Determining the nature of Chinese military policy and China's ongoing military buildup will be one of the 21st century's key international security issues. A wide variety of governmental, military, and scholarly assessments exist on the intentions and goals of China's military.[12] The secretive and dictatorial nature of China's government and military planning limit the amount of credible information about Chinese military doctrine and strategy that can be found in open source literature. China does not publicly publish a genuine English language counterpart to U.S. national military strategy documents. The lack of transparency in Chinese military policymaking has been noted by numerous sources, including the Defense Department's annual report to Congress on China's military power. It is believed that Chinese military doctrine places high emphasis on seizing the initiative in conflicts and keeping adversaries off balance through deception at the strategic, operational, and tactical levels. The absence of true transparency by the Chinese government and military about its military policies and doctrines, however, make any qualitatively reliable interpretation of Chinese military activities highly problematic.[13]

Since 1998, the Chinese government has biennially published what it says are English language national defense white papers, as well as selected papers on related topics such as Taiwan and its national space policy, at http://english.gov.cn/. The 2006 defense white paper (http://www.china.org.cn/english/feature/book/194421.htm) features sections on what China sees as the international security environment and China's role in that environment; a statement of national defense policy and organizational structure; descriptions of military force components such as the People's Liberation Army (PLA) and its border and costal defense program; declaration of how science and technology influence national defense strategy; purported defense expenditures; and appendices featuring major international exchanges between China's military and foreign militaries and the names (but not the text) of major military regulations issued during 2005–2006.

Interesting aspects of the document include the rhetorical declaration that China wants to build a society that is moderately prosperous, ethnically harmonious, stable, and making social progress. While contending that China's overall security environment is sound, it launches a diatribe against Taiwan for its purported desire to achieve national independence and its alleged threat to Chinese and Asian-Pacific regional security. The document also expresses its hyperbolic rhetorical concern over U.S. weapons sales to Taiwan and how some countries have created a "China threat."[14]

In terms of strategic defense doctrine, this treatise reveals that China places high emphasis on the important role of information technology and mechanization as driving forces in developing the PLA. It also stresses China's need to improve its national firepower, assault, mobility, protection, and information capabilities; enhance its efforts to build a joint operational system capable of fighting information based wars; move from a local defense posture to one capable of engaging in regional power projection; and retain a nuclear deterrent capable of deterring hostile powers.[15]

Additional credible, English-language information on Chinese military doctrine available through Chinese government or military Web sites is limited. The Central Military Commission, which is the organization responsible for commanding Chinese military forces, has miniscule English language content at http://english.gov.cn/2008–03/16/content_921750.htm. There is no English language Web site for the Ministry of National Defense, the PLA, or for Chinese professional military educational institutions such as the Academy of Military Sciences, which would be responsible for formulating the intellectual foundations buttressing Chinese military doctrine.

Learning more about Chinese military doctrine requires using open source resources and analyses produced by western governments and think tanks. Examples of some of these resources include National Defense University Library's Military Policy Awareness Links (MIPALS) (http://merln.ndu.edu/index.cfm?type=page&pageID=3), National Defense University's Center for the Study of Chinese Military Affairs (http://www.ndu.edu/inss/China_Center/INSS_About_CSCMA.htm), the Congressional-Executive Commission on China (http://www.cecc.gov/), the U.S.-China Economic and Security Review Commission (http://www.uscc.gov/), the U.S. Army War College's Strategic Studies Institute (http://www.strategicstudiesinstitute.army.mil/), the U.S. Army's Foreign Military Studies Office (http://www.leavenworth.army.mil/fmso/), the Project on Defense Alternatives China Military Power site (http://www.comw.org/cmp/), and the China military section of globalsecurity.org accessible at http://www.globalsecurity.org/world/military/china/.

Estonia

Estonia's complicated history, which includes its forcible annexation by the Soviet Union from 1940–1991, influences its current foreign and security policies, as does its location at the eastern end of the Baltic Sea, which requires it to be in ongoing consultation about economic, political, and security issues with other Baltic countries and the Russian Federation.[16] Given its vulnerability to territorial ambitions, Estonia has sought to maximize its security since regaining independence in 1991. It has also sought to minimize complications brought about by the legacy of Soviet occupation by seeking to join the European Union (EU) and North Atlantic Treaty Organization (NATO), which it succeeded in joining in 2004.[17]

In 2004 Estonia prepared and released its official military policy document, *National Security Policy Concept of the Republic of Estonia*. Accessible at http://merln.ndu.edu/whitepapers/Estonia-2004.pdf, this document emphasizes that Estonian national security policy is predicated on its membership in NATO and EU and upon defending common democratic values. It goes on to assert that Estonia will actively work with NATO and the EU to improve member state cooperation, that it will participate in the international security system according to its national commitments and capabilities, and that it will develop its national military defense in cooperation with allied countries.[18]

This document goes on to stress that the most serious threats to Estonia's security are possible instability, uncontrollable developments, and international crises, asserting that NATO and EU enlargement has significantly increased coverage of the European stability and security zone. Examples of possible incidents that this document says could threaten Estonia include increasing or unexpected military force deployments near Estonia's borders; large-scale military maneuvers near the country's borders that do not adhere to international arms control treaties; intentional violations of national air space, land, or waters; transport, radiation, or chemical accidents with cross-border repercussions; natural resource depletion; Estonia's acute dependence on foreign electricity and gas supplies; and computer crime.[19]

A number of resources document and analyze Estonian military policy and doctrine. These include the Estonian Ministry of Defence (http://www.kmin.ee/), whose publications include *Baltic Defence Cooperation* (2002), *Estonia and International Peace Operations* (2002), and *Estonia Defense Forces 2003–2006* (2002?), and the Estonian Defence Forces (http://www.mil.ee/).

An additional useful resource is the Baltic Defense College (http://www.bdcol.ee/) and the information resources produced by this professional military educational institution. Examples of these resources include the scholarly journal, *Baltic Security and Defense Review*, and its predecessor, *Baltic Defense Review* (1999–present).

Finland

Finnish military doctrine has been historically influenced by that country's location in the northeastern Baltic between Germany and Russia and the need to preserve its national sovereignty since its modern national independence only dates from 1917. A critically important and controversial component of Finnish national military strategic document and 20th-century foreign policy was Finlandization.

This policy kept Finland from being aligned with NATO or the European Union for much of the Cold War period, from approximately 1945–1991. Finnish President Urho Kekkonen (1900–1996), who served as President from 1956–1981, is considered the chief promulgator of this policy, which effectively saw Finland align its foreign policy and national security interests with those of the former Soviet Union, while retaining domestic political freedom and a modicum of foreign policy autonomy in other areas of the world. Finlandization also had its origins in the 1939–1940 Finnish War and the 1948 Finnish-Soviet Treaty of Friendship, Cooperation, and Mutual Defense, and its defenders would seek to rationalize this policy as being motivated by the geopolitical strategic necessity. Its critics, however, emphasize that such subservient behavior toward the Soviet Union during the Cold War era reflected poor moral judgment by a country proclaiming to adhere to democratic values.[20]

Finland was expected to gravitate toward the West after the Soviet Union's collapse, but while it did join the EU in 1995,[21] it has not joined NATO. A 1995 article by the Commander-in-Chief of the Finnish Defence Forces stated that nonalignment is Finland's best way to preserve northern European stability; however, he admitted that this situation could change if necessary. He went on to add that Finland was capable of mobilizing a force of over 500,000 personnel on short notice to defend its national territory and that the key component of Finnish military doctrine was creating a territorial defense system to wear down and delay invading forces with concentrated firepower.[22]

Additional Finnish post–Cold War security concerns included integrating the former Baltic states into Europe in ways similar to Finland's policy of avoiding provocation with Russia; ensuring that NATO expansion does not make Finland a front-line state in a potential confrontation with Russia; and maintaining Finland's independent defense capabilities.[23]

Finland's most recent military doctrine and strategy statement was published in 2004 by the Prime Minister's Office and the Ministry of Defence, and is accessible at http://merln.ndu.edu/whitepapers/Finland_English-2004.pdf and through the Defence Ministry Web site (http://www.defmin.fi/english/). *Finnish Security and Defence Policy* emphasizes Finland's desire to cooperate with the European Union's Common Foreign and Security Policy, NATO's Partnership for Peace, the Organization for Security and Cooperation in Europe, and the Council of Europe in enhancing European security architectures. It goes on to add that Finland engages in active and comprehensive conflict prevention and crisis management policies; that it is developing sufficiently trained and equipped forces that can be quickly deployed to international crises areas; that it will actively participate in international efforts to prevent proliferation of mass destruction weapons; that it will contribute actively to improving EU counterterrorism policies; and that it will also seek to prevent and combat environmental threats as they may affect shipping in the Baltic Sea and Gulf of Finland.[24]

This document's summary goes on to add that Finland seeks to develop its defense assets as a militarily unaligned country; that it is particularly attentive to changes in the Northern European security environment; that it uses conscription and a territorial defense system as the basis for defending the entire country; that its forces are prepared to prevent and repel hostile attack; that it is expanding its Army's readiness brigades' firepower and mobility; enhancing the Army's ground-based and regional forces; increasing the Navy's ability to protect sea lines of communication and develop mobile coastal troops; and enhancing the Air Force's fighter defense assets and air defense command and control system.[25]

A 2006 follow-up document outlines Finnish national defense strategy until 2025. Salient points of this report discuss factors that could affect Finland's national security environment, including a declining Russian population, aging European populations, and increasing populations of developing countries; Finland's dependence on energy imports and natural resource scarcity, including uneven international food distribution quality; and that price increases for new technology and increasing global economic interdependence may also drive international conflict. In addition, this report stresses that the military conflict spectrum will expand with traditional boundaries between war and peace, becoming more muddled, and that asymmetric warfare will be increasingly common; that international crises will require earlier intervention from greater geographic distance; and that the Baltic Sea's importance to Russia will increase due to critical energy and material transportation.[26]

Further resources for Finnish military doctrine include other documents on the Ministry of Defence Web site, the Finnish military Web site (http://www.mil.fi/) (although it lacks English language content), the Finnish Institute of International Affairs (http://www.upi-fiia.fi/eng/), the previously mentioned Baltic Defense College, and Finland's National Defense University (http://www.mpkk.fi/en/), which includes some English language analysis of national security issues produced by entities such as the Department of Defense and Strategic Studies, including reports such as *EU Battlegroups: Theory and Development in the Light of Finnish-Swedish Cooperation* (2005).

France

French military history and doctrine have been influenced by multiple factors. These include the lofty ambitions of the Napoleonic era, an extensive colonial empire in regions such as Africa and the South Pacific, which has given France global security ambitions and interests, the trauma of defeat and occupation during World War II, the development of a nuclear deterrent during the presidency of Charles De Gaulle, the desire to remain independent of the United States by withdrawing from NATO, the desire to play a leading role in developing European Union security policy, and the need to develop strategies to combat Islamist terror in areas such as Afghanistan and within French territory.[27]

The most recent statement of French military doctrine is its white paper on defense and security, which was issued in June 2008 to update previous documents from 1972 and 1994.[28] Highlights of this document include concerns about jihadi-inspired terrorism aiming directly at France and Europe; France becoming more vulnerable to ballistic missiles developed by powers such as Iran; French security priorities needing to concentrate on an arc of vulnerability encompassing the Atlantic, Mediterranean, Arabian-Persian Gulf, and Indian Ocean; the need for France to have freedom of action to conduct operations in various African theaters, including the Sahel; maintaining an effective and diversified nuclear deterrent capability; making the European Union a major player in European crisis management and international security by having an intervention capability of 60,000 soldiers deployable for one year in a distant theater; stressing the complementary nature of the European Union and NATO; and advocating full French participation in NATO structures.[29]

To respond effectively to these security issues with the appropriate force structure, the White Paper makes a number of recommendations that must receive French parliamentary approval. These include equipment modernization, with particular emphasis on force and equipment protection, intelligence, and information security; maintaining an aircraft carrier group; having a joint fleet of 300 combat aircraft; increasing defense spending one percent a year above pension spending between 2012–2020; increasing European defense industry integration without compromising French nuclear force and cyber-security capabilities; maintaining the highest possible professional standards for military and civilian support personnel; and doubling funding for satellite programs and establishing a Joint Space Command.[30]

The French Army's Centre du Doctrine d'Emploi des Forces (CDEF) (http://www.cdef.terre.defense.gouv.fr/) has a number of resources in French and English that describe and analyze French military doctrine. These include reports such as *Ongoing Reflections on the Future Employment of Land Forces* (2005) and *Multinational Operations and Forces Command: French Commanders* (2007) and articles from the journal *Doctrine* (December 2003–present), with representative samples including "The Contribution of the Armed Forces in the Stabilization Processes," no. 12, August 2007, and "UAV-Helicopter Co-operation: A Promising Course of Action," no. 14, January 2008.

A particularly significant CDEF publication is *Winning the Battle Building Peace: Land Forces in Present and Future Conflicts FT-01* (2007). This document describes the increasing importance of asymmetric conflict in conducting military operations and emphasizes how this has changed the role of military operations and soldiers participating in these operations. It emphasizes the importance of cooperation with local populations and the importance of working with these populations to conduct such operations and achieve peace following the conflict.[31]

Particular importance is placed on stabilization in military operations as the following excerpt demonstrates:

> The stabilisation phase is the decisive phase of a military operation; the decisive action is carried out on the ground, at the heart of human society. It is here that armed forces establish the conditions for strategic success. The

> stabilization phase depends to a large extent on a preparation which, involving numerous actors, starts with the concept of the operation, and allows for a successful transition from one phase to another as this profoundly influences the future course of the conflict. The success or failure of the stabilization phase is often determined by the beginnings.[32]

Additional sources listing and analyzing French military doctrine (with these being predominately in French) include the Ministry of Defense (http://www.defense.gouv.fr/), the St. Cyr military academy (http://www.st-cyr.defense.gouv.fr/), the Naval School (http://wwwold.ecole-naval.fr/), the Defense College (College Interarmees de Defense) (http://www.college.interarmees.defense.gouv.fr/), Centre des Interarmees de Concepts, de Doctrines et d'Experimentations (http://www.cicde.defense.gouv.fr/), Delegation aux Affaires Strategiques (http://www.defense.gouv.fr/das/), Center for Prospective and Strategic Studies (http://www.cerens.defense.gouv.fr/), and Fondation pour la Recherche Strategique (http://www.frstrategie.org/).

Germany

German military forces have played an important historical role in developing national military doctrine, and significant literature documents how this doctrine has influenced German national security policy and the military policies of other countries.[33] One of the most famous and controversial examples of German contributions to military doctrinal thought was the Schlieffen Plan formulated for World War I by Field Marshall Alfred Count von Schlieffen (1833–1913). Schlieffen's military plan for a potential European conflict called for Germany to fight a two-front war with France and Russia by placing primary emphasis on defeating French forces in the west by passing through neutral Belgium before using Germany's superb railway network to transport these forces to the east to defeat Russia.[34]

Germany's ultimate defeat in World War I and the harsh terms of the Versailles Peace Treaty sent German military planners back to the drawing board. The interwar years saw covert cooperation with the Soviet Union and the development of the military doctrine of blitzkrieg, which would be used with considerable success during World War II's opening campaigns. Germany's allied opponents would eventually stymie and reverse the German successes at high cost and the Wehrmacht's initial invincibility would be reversed, causing this once indomitable force to experience a more complete defeat than in World War I and end the policymaking and strategic environment that allowed such military doctrine to develop.[35]

Following Germany's defeats in both World Wars, the development of a unique national military doctrine took a backseat to national planning, as a divided Germany became part of NATO and Warsaw Pact military force planning between 1945 and 1990. This period also saw antimilitarism increase within national political discourse as a result of these defeats.[36] The collapse of the Soviet Union and

Warsaw Pact between 1989 and 1991 set in motion a process that would result in German reunification in 1990.[37]

These epochal events would, in turn, drastically alter Germany's national security situation. A crucial factor to resolve would be the withdrawal of Russian troops from the former East Germany, which was accomplished by 1994.[38] Germany would spend the next few years trying to absorb the former East Germany, and this timeframe would also see the tentative emergence of a debate within German security circles over what military role Germany should play in the post–Cold War world. Some of this debate would be ignited by turmoil in the former Yugoslavia, while the events of 9/11 and afterward would cause German policy-makers to explore the possibility of German military operations outside of NATO or EU frameworks.[39]

The reunified German government would issue its first military doctrine document in 1994. This white paper acknowledged NATO's drastic reductions in its nuclear arsenal and withdrawal of ground-launched short-range nuclear weapons; mentioned Germany achieving unity with the approval of its neighbors and world powers while remaining in NATO; acknowledged that Germany must assume new international security responsibility; recognized that Germany now played a central role in furthering European integration and enhancing the transatlantic partnership and the United Nations; understood that unstable regions in Europe, Asia, and Africa increased international security uncertainty; and acknowledged that traditional concepts of deterrence and defense were not suitable to resolving domestic and social conflicts.[40]

The 1994 White Paper went on to assert that to meet emerging security challenges, the German military (Bundeswehr) must have reconnaissance assets capable of detecting threats to Germany and NATO in a timely manner; that Bundeswehr and allied land forces would need to be able to protect Germany from an attack against German territory; that its air forces needed to be capable of conducting peacetime air surveillance operations and conduct wartime defensive and deep battle support operations with allies; and that naval and naval air forces would need to work with allies to keep open communication sea lines and prevent enemy landings on German soil.[41]

In 2003, Germany's Defense Ministry issued *Defense Policy Guidelines*. Highlights of this document included emphasizing the vital importance of the transatlantic partnership to German security; that Germany will only conduct military operations with UN, NATO, and EU allies and partners, with the possible exception of evacuation and rescue missions; that its armed forces are integrated into NATO more than any other ally; and that current and future Bundeswehr operations require it to be capable of participating in multinational operations across the combat spectrum and outside allied territorial boundaries.[42]

Subsequent years would see German military forces accelerate their efforts to achieve greater technological capabilities[43] and send troops to conduct combat operations in Afghanistan as part of NATO's International Security Assistance force, although the effectiveness of these German troops has been questioned due to restrictive rules of engagement.[44]

Germany's most recent defense white paper was issued by the Ministry of Defense in 2006. This document strongly stresses the important role that terrorism and weapons of mass destruction play in German domestic and international military policy doctrine and indicates an apparent willingness to play a more proactive role in dealing with these threats, as the following excerpt demonstrates:

> International terrorism represents a fundamental challenge and threat to freedom and security. Increasingly, the proliferation of weapons of mass destruction and of the means of their delivery has become a potential threat to Germany as well as to other nations. In addition, Germany has been confronted with the aftermath of intrastate and regional conflicts, the destabilisation, and the internal disintegration of states as well as its frequent by-product—the privatization of force. Strategies that were previously effective in warding off external dangers are no longer adequate against the current, asymmetric threats. Today's security policy must address new and increasingly complex challenges. Effective security provisions require preventative, efficient, and coherent cooperation at both the national and international levels, to include an effective fight against the root causes. It is imperative that we take preventive actions against any risks and threats to our security and that we address them in a timely manner and at their sources.[45]

The 2006 German defense white paper also stresses the important role of conscription in sustaining the Bundeswehr and that it will continue; that Germany will continue relying on indigenous and international defense industrial technological capabilities to enhance national security policy; and the Bundeswehr's commitment to enhancing its ability to operate in a multinational environment. It also features a chart showing German participation in various international peacekeeping missions.[46]

Numerous military and civilian resources list and analyze contemporary military doctrines and policy in German or English. These include the German Defense Ministry (http://www.bmvg.de/); the Bundeswehr (http://www.bundeswehr.de/); Helmut Schmidt Universitat-Universitat der Bundeswehr-Hamburg (http://www.hsu-hh.de/hsu/) and Universitat der Bundeswehr-Munich (http://www.unibw.de/), which serve as Germany's premier professional military educational institutions; the George C. Marshall Center European Center for Security Studies (http://www.marshallcenter.org/); Bundesakademie fur Sicherheitspolitik (http://www.baks.bundeswehr.de/); and the Stiftung Wissenschaft und Politik: Deutsches Institute fur Internationale Politik und Sicherheit (German Institute for International Politics and Security) (http://www.swp-berlin.org/).

India

During its six decades of independence, India has gone from a poor, developing country to an increasingly important factor in Asian security policymaking. India is a recognized nuclear weapons producing state, has expressed an interest in developing a military space program, and faces a diverse variety of national security challenges. These security challenges include its complicated and tense relationship with Pakistan, which is exacerbated by that country's support of Kashmiri separatists, and India's geographic proximity to countries with unstable political regimes and security conflicts, such as China, Afghanistan, Iran, Myanmar, Nepal, and Sri Lanka. These matters have also influenced the development of Indian military doctrine, the scope of which covers conventional and nuclear forces. A continually growing corpus of scholarly literature reviews and analyzes these security challenges and Indian military doctrine.[47]

Annual reports produced by India's Ministry of Defence provide information on how that country views its international security environment. The 2007/2008 *Annual Report* notes that global attention is shifting to the Indian subcontinent for reasons such as its accelerated economic growth, growing population and markets, and increasing energy consumption. This document also describes Pakistan's deteriorating situation, symbolized by former Prime Minister Benazir Bhutto's assassination, continuing unrest in Afghanistan and Sri Lanka, and the need for peace in the Persian Gulf region where several million Indian nationals live and which is the key source of India's energy supplies, as particularly important to Indian national security interests.[48]

This document further stressed that India seeks to follow a policy of constructive engagement with China; that there is no military solution to the internecine conflict in Sri Lanka; India's support for making Afghanistan a viable democratic state; India's desire to support political reform in Myanmar; its concern over the role of international terrorism and its contention that effective counter infiltration operations along the Line of Control has reduced terrorist attacks in Jammu and Kashmir; and its desire to maintain a strong defense force to increase growth, stability, and peace and its preparation to deter conventional and unconventional military threats.[49]

Following its 1998 nuclear explosions, India began working on developing doctrine for its nascent nuclear arsenal. A draft report released in August 1999 by the National Security Advisory Board on Indian Nuclear Doctrine stressed that India would pursue a policy of "credible minimum deterrence," that India's nuclear policy would be retaliation only, and that an effective deterrent required India to maintain sufficient:

- Survivable, and operationally prepared nuclear forces;
- A robust command and control system; effective intelligence and early warning capabilities;
- Comprehensive planning and training for operations to align with this strategy; and
- The will to use nuclear forces and weapons.[50]

In January 2003, India finalized its nuclear command structure and formalized its nuclear doctrine. This doctrine was crafted as a result of debate within the Indian defense and military establishments, reaction from the United States, China, and Pakistan, and regional security developments with Pakistan such as

Operation Parakram, a 2002 Indian Army deployment along the Pakistani border. Attributes of Indian nuclear doctrine, which were refined slightly from the 1999 draft version, included:

- Building and maintaining a credible minimum deterrent;
- Adopting a no first use nuclear weapons policy;
- Ensuring that retaliatory military attacks can only be authorized by civilian political leadership through the National Command Authority;
- Not using nuclear weapons against nonnuclear weapons states;
- Retaining the option to retaliate with nuclear weapons if India or Indian forces are attacked by biological or chemical weapons;
- Continuing export controls on nuclear and missile-related materials and technologies; participating in Fissile Material Cut-Off Treaty negotiations; observing a nuclear testing moratorium; and working toward nuclear disarmament.[51]

India's most recent military doctrine document was released by its Army in October 2004. This document stressed an increasing emphasis on maneuver and jointness between its armed services branches and a particular emphasis on information warfare and network-centric warfare. This document accented the need to reduce and improve the military's wartime decision making and disrupt the enemy's decision cycle, which it says was a hallmark characteristic of U.S. campaigns from the 1991 Persian Gulf War to the conventional phases of 2003's Operation Iraqi Freedom. Additional attributes of this doctrine include the importance of India and Pakistan avoiding a military confrontation to prevent a nuclear war from occurring, and recognition that while network-centric warfare may increase the uncertainty of enemy decision making, it may also have the side effect of producing greater confusion and leading these opponents to make errors in judgment, which could produce unplanned conflict escalation.[52]

Numerous resources may be used to consult Indian military doctrine documents and analysis of this literature. The annual reports section of India's Ministry of Defence Web site (http://mod.nic.in/) provides documents from 1999–2000 to the present. India's Army Web site is http://indianarmy.nic.in/, the Navy's Web site is http://indiannavy.nic.in/, and the Air Force's Web site is http://indian airforce.nic.in/. Additional useful military-related Web sites include the Defence Services Staff College (http://armedforces.nic.in/interservice/isidssc1.htm) and the National Defence College (http://ndc.nic.in/).

Civilian policy research organizations that analyze Indian military doctrine include the Institute of Peace and Conflict Studies (http://www.icps.org/), the Institute for Defence Studies and Analyses (http://www.idsa.in/), and the Strategic Foresight Group (http://www.strategicforesight.com/).

Indonesia

Indonesian military policy and doctrine have developed over the six-decade period since it gained independence in 1945. This southeast Asian island archipelago nation's boundaries are located at the approximate intersection of the Indian Ocean, South China Sea, and Pacific Ocean, with the critically important international trade corridor of the Strait of Malacca being within Indonesian territorial parameters. The military has played an important and controversial role in recent Indonesian political history, with Indonesia's occupation of East Timor from 1975–1999 representing the most vivid and contentious international example of Indonesian military activity. Military dictatorships played a dominant role in Indonesian political history until revolutions in 1998 led to a gradual reduction of the military's preeminence in Indonesian political life.[53]

Indonesian military doctrine from approximately 1945–1998 evolved from its independence struggle against the Dutch. Called "Total People's Defense and Security," it emphasized guerrilla warfare that involved support and assistance from the civilian population and merged civilian and military cadres. Since the 1998 revolution, a new doctrine called "New Paradigm" has been implemented. New Paradigm was developed by senior officers, such as Lieutenant General Susilo Bambang Yudhoyano, who believed Indonesian armed forces (TNI) needed to change to accommodate Indonesian societal changes.[54]

Critical elements of New Paradigm see TNI's traditional focus shifting from internal security to external defense. The national police, originally under TNI command, were established as a separate organization reporting to the President and were given responsibility for internal security functions. New Paradigm requires the police to develop paramilitary capabilities to deal with insurgencies and large internal security threats. TNI can only assist the police if the police are unable to handle a situation and if TNI is directed to by central authorities. In 2001, Indonesia's parliament recognized that only the Army had the capability to ensure public order and confront armed separatist movements when it passed legislation that assigned TNI four internal security missions, including operations against separatists, insurgent forces, drug trafficking, and smuggling.[55]

Indonesia currently has no significant conventional external threat to its national security other than international terrorism. The regional terrorist group Jemaah Islamiyah conducted bombing attacks against night clubs in Bali in October 2002 and the Jakarta Marriott Hotel in August 2003, which achieved significant fatalities, as did a bombing of the Australian Embassy in Jakarta in September 2004. Cooperative Indonesian and international investigation of these assaults produced a number of arrests and revealed a network of terrorist cooperation whose membership included Al Qaida. In addition, piracy in the Strait of Malacca, smuggling, and maritime poaching also threaten Indonesian security.[56]

Indonesia also faces a number of internal security threats stemming from terrorism and ethnic and religious conflict. Besides Jemaah Islamiyah, there are separatists in Aceh and Papua. The Gerakan Aceh Merdeka (GAM-Free Aceh Movement) seeks an independent Islamic state in Aceh, and Organisasi Papua Merdeka (OPM-Free West Papua Movement) seeks independence for Papua. There have been religious violence incidents in Maluku and Central Sulawesi, ethnic violence over land use in Kalimantan and other areas, incidents of anti-Chinese riots in urban areas; and instances of radical Muslim groups threatening westerners in tourist areas and cities such as Jakarta.[57]

Indonesia's most recent military policy statement was issued in 2003. *Defending the County in the 21st Century* covers topics such as recent national political and defense reform; Indonesia's domestic and international strategic context; how Indonesia will use its defense forces to defeat traditional and non-traditional security threats; and its defense cooperation with countries such as Australia, China, Malaysia, Singapore, and the United States.[58]

This document goes on to stress the increasing importance of Military Operations Other Than War (MOOTW) in TNI doctrinal activity. Examples of such activities include counterterrorism operations; battling separatist groups in Aceh and Papua; fighting piracy and illegal immigration; resolving communal disputes; overcoming illegal fishing, logging, and other environmentally destructive activities; assisting civil governments in mitigating natural disaster impacts; providing search and rescue assistance; and participating in international peacekeeping operations. This white paper also expressed concern that its ability to meet these obligations was impeded by national budget restrictions giving defense spending only one percent of Gross Domestic Product (GDP) with the regional Southeast Asian national defense expenditures averaging over two percent of GDP.[59]

Additional insights on Indonesian military doctrine can be gained from Indonesia's Ministry of Defense (http://www.dephan.go.id/), the TNI (http://www.tni.mil.id/), the Indonesian Army (http://www.tniadmil.id/), and Indonesia's National Resilience Institute (Lemhannas) (http://www.lemhannas.go.id/). However, these sites are in Indonesian. The S. Rajaratham School of International Defense Studies at Singapore's Nanyang Technological University (http://www.idss.edu.sg/) provides helpful English language insight on southeast Asian security issues, which can include analysis of Indonesian military matters. Additional English language resources include the Australian Strategic Policy Institute, Australia's Lowy Institute for International Policy (http://www.lowyinstitute.org/), and the Australian National University's Strategic and Defense Studies Centre (http://rspas.anu.edu.au/sdsc/), and the Centre for Strategic and International Studies (http://www.csis.or.id/).

Israel

During its six decades of modern existence, Israel has had to contend with a hostile national security environment in which most of its surrounding neighbors have sought to destroy it. While Israel has achieved some semblance of peace with Egypt and Jordan, it still confronts a hostile security environment with threats from Iran, instability in Lebanon, Palestinian terrorists seeking to derail a fragile Palestinian state and an Israeli-Palestinian peace process, and Syria. Israel has been forced to fight four major wars and several localized and often ongoing conflicts to ensure its physical survival and maintain its national security interests. Consequently, Israel has been forced to develop highly diversified capabilities to meet its multifaceted national security requirements and to develop a variety of doctrinal tactics for addressing these military exigencies. Significant literature on Israeli military doctrine provides historical and contemporary analysis on how this doctrine has been structured and its overall effectiveness.[60]

Formulators of Israeli military doctrine have had to address topics such as conducting conventional land operations with armor and artillery; aerial operations against national armies like Egypt and Syria; counterterrorism and counterinsurgency operations against Palestinian forces; and nuclear doctrine for its own nuclear arsenal and to counter potential nuclear threats from countries like Iran. The three primary pillars of Israeli military strategy have been deterrence, strategic warning, and decision. Since the early 1950s, Israel has sought to maintain the status quo by using military threats to deter rivals by threatening serious punishment if Israeli defenses are challenged. The strategic warning component of Israeli deterrence has included building up conventional and, arguably, nuclear capabilities and demonstrating the resolve to use these assets against adversaries. Israel's ability to rapidly mobilize its reserve forces is a critical factor in demonstrating national military resolve. If war begins, Israel seeks to defeat its opponents decisively and swiftly, with the Six Day War of 1967 serving as the most vivid demonstration of this, since Israel doesn't have the manpower resources to conduct prolonged military conflicts. Punishment is also part of Israeli deterrence strategy, with occupation of Arab territories being a critical bargaining operational goal for future diplomatic negotiations.[61]

A 1991 analysis of Israeli military doctrine published in the *IDF Journal* asserted the importance of achieving victory in the shortest possible time, relying on surprise and acquiring new weapons systems, restructuring, and appropriate strategy as traditional components of Israeli military doctrine. This assessment went on to maintain that Israel needed all of these attributes and increased defense spending to counter changes in the emerging regional strategic environment, such as the growing economic purchasing power of neighboring nations that makes it possible for them to purchase precision high technology weapons to threaten Israeli military strengths.[62]

Additional attributes of Israeli military doctrine include the high levels of responsibility and freedom of action given to junior officers:

- Entrusting junior leaders with generous amounts of initiative and stating that the leader closest to the battle has the best knowledge of what is going on and should be the decision maker;
- Deemphasizing "spit and polish" discipline;
- Maintaining high military proficiency with an acute stress on tough realistic military training and fighting discipline;
- Developing close relationships between junior officers and subordinates;
- Officers setting an example by providing leadership from the front, including sacrificing their own needs for the safety and comfort of their subordinates.[63]

The early years of the 21st century have seen a defensive homeland security capability and precision-guided munitions added to Israel's reliance on the deterrent power of its offensively oriented military doctrine. This change in Israeli military doctrine has been bolstered by the recognition that the threats Israel faces are not deterred by this traditional deterrent mechanism. Key components of the now-prevalent threat to Israel include:

1. Terrorist disruption of Israel's economy and society that could isolate Israel diplomatically and strategically if Israel's responses are viewed as disproportionate.
2. Ballistic missile strategic attacks from Iran, Syria, and Lebanon and potentially other countries.
3. International political pressures, like those described in Point 1, that limit Israel's ability to make sound and independent military judgments, which may severely limit or damage Israeli security and prosperity.[64]

There is not a single publicly accessible Israeli government Web site with the text of Israeli military doctrine or national security policy. Some information and analysis of Israeli military doctrine can be found on a selection of Israeli government Web sites, including the Ministry of Defense (http://www.mdf.gov.il/), Israel Defense Forces National Defense College (http://dover.idf.il/IDF/English/units/other/pum/Background.htm), and Israel Defense Forces (http://www.idf.il/). The Israeli Ministry of Foreign Affairs (http://www.mfa.gov.il/) has an English language summary of the Winograd Commission report documenting military failures in the 2006 Hezbollah War, but the full text of the report is only available in Hebrew.

Representative Israeli academic and public policy research institutions analyzing Israeli military doctrine include the Ariel Center for Policy Research (http://www.acpr.org/il/), Bar Ilan University's Begin-Sadat Center for Strategic Studies (http://www.besacenter.org/), Institute for Counterterrorism (http://www.ict.org.il/), National Security Studies Center (http://nssc.haifa.ac.il/), Tel Aviv University's Institute for National Security Studies (http://www.inss.org.il/), and the University of Haifa's Reuven Chair in Geostrategy (http://geo.haifa.ac.il/~ch-strategy/).

Russia

Russian military doctrine has received extensive historical and contemporary analysis of the Soviet era from 1917–1991, with somewhat less emphasis placed on the post–Communist era. The Russian military has been increasingly assertiveness under the nationalistic leadership of Vladimir Putin, as evidenced by its August 2008 invasion and occupation of the Abkhazia and South Ossetia regions of Georgia, and it has been augmented by increased oil and natural gas revenues, which have enabled Russia to devote more financial resources to its military, including its nuclear forces. The Soviet military had an extensive corpus of military doctrine for conventional and nuclear forces. Some of that doctrine has been retained by the Russian Federation, while portions of it have been updated to accommodate existing and emerging strategic realities in accord with what Russian national security policymakers consider as vital national interests. There is also significant literature documenting and analyzing Soviet and Russian military doctrine.[65]

The collapse of the Soviet Union between 1989 and 1991 drastically reduced the territorial size of the Russian Federation that emerged in the aftermath. The economic and political upheaval of these events also reduced the economic resources available to the Russian Government for its military. One of the first examples of post–Soviet Russian military doctrine was enunciated by Boris Yeltsin's government in 1993. This document envisioned that Russia would have no potential enemies, while calling on its military to develop so it could defend itself and the Russian people. Operational attributes of this document, which were in contrast to a more defensive posture adopted by the Soviet military during the Gorbachev era, include:

- Changing from a defensive position to having a preemptive strike capability.
- Reverting from proclaiming no nuclear weapons use to envisioning possible escalatory nuclear weapons use.
- Putting increasing emphasis on strategic non-nuclear forces, such as Submarine Launched Ballistic Missiles, ICBMs, and air and sea-launched cruise missiles.
- Placing new emphasis on military technology advances in command, control, communications, computers, and intelligence (C4I), long-range smart weapons, and increased air and space mobility.
- Announcing a willingness to retaliate in response to "hostile" action taken against ethnic Russians living in former Soviet states.[66]

The next evolution in Russian military doctrine was released on April 21, 2000, with the imprimatur of new Russian President Vladimir Putin. This document asserted that key features of modern war included its coalition nature and its affect on all areas of human activity; the extensive use of indirect and non-traditional combat operations, including electronic engagement; both participants desiring to disrupt governmental and military command and control systems; using highly maneuverable operational forces, such as airborne troops and special forces; attacking rear-service economic facilities and the opponents' communication assets; the serious consequences of hitting and destroying power-generating, chemical, and other critical infrastructure facilities; the increasing possibility of new states being drawn into war; the possible use of weapons of mass destruction; and irregular armed forces participating in operations with regular military forces.[67]

This 2000 document went on to maintain that critical mission responsibilities of Russian Federation armed forces included:

- Responding in a timely matter to political or military threats to the Russian Federation and its allies;
- Maintaining combat and mobilization readiness for conventional and strategic nuclear forces;
- Protecting and defending national borders;
- Increasing air defense integration;

- Safeguarding information security and technical communication capabilities; and
- Conducting unilateral or multilateral strategic operations against hostile forces.[68]

A subsequent evolutionary update in Russian military doctrine was the October 2, 2003, issuance of *Urgent Tasks for the Development of the Armed Forces of the Russian Federation*. Called the Ivanov Doctrine, in honor of then–Minister of Defense Sergei Ivanov, this document examined the capabilities Russia needed to fight modern wars and discussed how to enhance its power projection capabilities. Viewpoints expressed in the Ivanov Doctrine are a composite of policy debates involving the Ministry of Defense and the military's General Staff, along with the political environment of the upcoming (December 2003) Duma legislative elections and Putin's March 2004 presidential reelection campaign.[69]

An additional factor influencing the Ivanov doctrine was the legacy of poor Russian military performance during the 1999–2001 Chechnya conflict. During these operations, Russian forces received poor advice from general staff planners, limited counterinsurgency training, were constricted by insufficient advanced reconnaissance and communications equipment, and possessed insufficient long-range precision guided munitions.[70]

The Ivanov Doctrine sought to respond to these problems and to U.S. operations in Afghanistan and Iraq by stressing Russia's commitment to transform its military into a force capable of countering diverse threats with fewer casualties and greater sophistication. This doctrine went on to emphasize the supremacy of ground forces and the need to enhance the ability of these forces to play a leading role in counterinsurgency operations. It also expressed that the Ministry of Defense would enhance individual combat standards and hire more professional, non-commissioned officers.[71]

Additional salient characteristics of the Ivanov Doctrine include shifting from combined arms operations to increasing the emphasis of air power as a policy instrument; promising to develop long-range precision-guided airborne missiles; and developing a lighter and more agile infantry force backed by improved strategic airlift capabilities. It also described the three primary goals of the Russian military transformation to include:

- Combating terrorism;
- Restoring national global power projection capability; and
- Consolidating Russian influence in the former Soviet Union.[72]

The increasing assertiveness of Russian foreign policy as demonstrated in Georgia, in recent attempts to assert territorial sovereignty in the Arctic, and the possibility of engaging in future military action against Ukraine or other former Soviet territories, means that keeping track of Russian military policy and doctrine documents will become increasingly important. There is no single English language source of Russian military doctrine documentation. Russian language sources that can be consulted include the Russian Military (http://www.mil.ru/), which also features some English language content, and the Russian National Security Council (http://www.scrf.gov.ru/).

Analysis of Russian military doctrinal documents can be found in English language translations of the Russian journal *Military Thought*, the *Journal of Slavic Military Studies* and other western military science and policy journals, the U.S. Army's Foreign Military Studies Office (http://fmso.leavenworth.army.mil/), the Army War College's Strategic Studies Institute (http://www.strategicstudiesinstitute.army.mil/), the Russian language site http://www.milparade.ru/, the International Institute for Strategic Studies (http://www.iiss.org/), the Carnegie Endowment for International Peace's Moscow Center (http://www.carnegie.ru/en/), and Russia's Institute for Strategic Studies (http://www.riss.ru/).

Singapore

As an island city-state in southeast Asia, Singapore is located on the strategic Strait of Malacca, which is an important international shipping point whose closure would have serious global economic impacts. Singapore's strategic importance was demonstrated during World War II when Japanese forces conquered this British colony, effectively ending an era of British colonial presence in Asia and paving the way for the postwar independence of many southeast Asian countries.[73]

Subsequent decades have seen Singapore rise from a third world county to an economically advanced and affluent nation-state that is an important factor in southeast Asian economic and security policymaking. Singapore has developed and continues to develop a small but highly skilled military capable of meeting many of its security needs. These requirements include keeping the Straits of Malacca open to international trade, combating terrorists or pirates that seek to jeopardize Singapore's access to its maritime surroundings, ensuring access to the natural resources Singapore must import to sustain its economic vitality, challenging relations with neighbors such as Indonesia and Malaya, with the latter country providing significant quantities of Singapore's water, and maintaining close security relations with countries as diverse as the United States, Australia, its partners in the Association of Southeast Asian Nations (ASEAN), and China.[74]

There is also concern over the possible consequences of losing access to the Straits of Malacca due to this body of water being blocked by a sunken tanker at the 1.5-mile wide Phillips Channel in the Singapore Strait. Such an incident is estimated to cost the global economy over $200 million annually and would require ships to be diverted further south, adding an extra one and a half days of sailing time. These concerns have lead Singapore to create a national policy organization to counter terrorism, adopting the following security measures:

- Requiring oil tankers to give 24-hour notice of their arrival and using high-tech identification systems to track their movements;
- Strengthening security at sea checkpoints such as the Singapore Cruise Center;
- Having Singapore Navy ships escort selected merchant vessels in territorial waters;

- Marking routes for ferries and other commercial vessels to keep them away from sensitive anchorages or installations; and
- Deploying radiation detection equipment at border entry points to screen containers and personnel for radiological materials.[75]

Singapore's most recent national security strategy document was published in 2004, and it placed counterterrorism as a critical element in its strategic policymaking. Emphasizing Southeast Asia's vulnerability to terrorist actions, the following introductory passage from this document stresses the long-term nature of the terrorist threat:

> Singapore is high on the list of targets for terrorist action. It is important that we recognize this reality. The extremist regional network Jemaah Islamiyah (JI), which is intent on subverting governments in the region, has targeted us before. These plots were foiled, but we can anticipate that there will be more attempts to attack us. Besides JI, we may face action from other extremist groups as well. Worldwide, Al-Qaeda elements remain active, planning future action against American and other interests. We are not alone in the struggle against terrorism. Yet, we must recognize that we are ultimately responsible for our own security. Terrorism is certainly not new to Singapore. It can be understood as the mounting of tactical operations aimed at achieving certain political goals. In terrorism, relatively little effort may be required to produce devastating results. It capitalizes on the element of surprise, but works over long time frames. Even if disrupted, terror organizations may regenerate themselves, and wait years before pursuing their objectives again.[76]

Singapore has sought to cope with these and other security threats by developing a Homefront Crisis Ministerial Committee and Homefront Crisis Executive Group to provide strategic and political crisis handling guidance and provide policy guidance and strategic decisions for managing major crises.[77]

It has also sought to bolster its military capabilities and doctrine by moving from a deterrence-based posture to a more expeditionary approach against enemies in order to ensure swift and decisive victory. This has involved its Army seeking to enhance its precision strike and networking capabilities and developing new urban fighting doctrine equipment and capabilities; its Air Force participating in the United States' Joint Strike Fighter Program and enhancing a multi-spectrum air defense capability; its Navy strengthening its collaboration and information sharing with regional partners to track ship movements and incidents; and the 2007 passage of legislation giving the military the legal authority to conduct operations supporting civilian law enforcement agencies.[78]

Numerous resources provide information on Singaporean military policy and doctrine, including the Ministry of Defence (http://www.mindef.gov.sg/), Singapore Armed Forces Technology Institute (http://www.mindef.gov.sg/safti/ and its scholarly journal *Pointer,* which is available online from 1998–present, and the

Ministry of Defence's Innovation and Transformation Office (http://www.mindef.gov/sg/innovation/) and Centre for Military Experimentation (http://www.mindef.gov.sg/fsd/scme/). A useful non-governmental information resource analyzing Singaporean military policy is the S. Rajaratnam School of International Studies at Nanyang Technological University (http://www.rsis.edu.sg/).

South Africa

Representing Africa's most prosperous and militarily powerful nation, South Africa has significant military capabilities that must be incorporated into any evaluation of African military matters. During the past six decades, South Africa's military has engaged in operations in locales as diverse as Angola, Lesotho, Mozambique, and Namibia, developed a nuclear weapons program that was eventually dismantled due to international pressure, let itself be stained by apartheid, which limited its human capital potential, endeavored to make the transition from apartheid by incorporating previously excluded individuals from its ranks, while retaining optimum professional standards, experienced defense spending cuts since the 1994 end of apartheid, developed and attempts to sustain a significant international arms export program, has had to contend with the prospect of becoming involved in domestic law enforcement due to South Africa's high crime rate in many areas, has defended a maritime area encompassing the intersection of the Atlantic and Indian Oceans, and may face the prospect of intervening militarily in Zimbabwe to oust the dictatorial regime of Robert Mugabe and the deteriorating humanitarian and security situation in that country.[79]

South African military doctrine has been influenced by historical factors such as its sometimes contentious relationship with Great Britain, the cultural consciousness and experiences of its Dutch-descended Afrikaner population, its relationships with indigenous and neighboring African peoples, and through conflicts such as the Boer War, two World Wars, and campaigns to defend apartheid policies through internal security operations or operations against neighboring countries seen as hostile to South African national security and political interests. The late 1960s saw the South African National Defence Force (SANDF) begin formulating the doctrine that it needed three divisions to carry out its military requirements. One of these divisions would be armored in order to destroy the enemy; a second division would be mechanized in order to maneuver around the enemy; and a third division would be infantry, which would take up geographic defensive positions to block, hold, or fix the enemy.[80]

An enemy strong enough to carry out such an invasion never materialized, and SANDF land forces never received these three divisions. Following the collapse of apartheid and the 1994 election of the African National Congress government, the reconstituted South African military had to begin integrating revolutionary anti-apartheid forces into its personnel, which was a task that frequently proved challenging for personnel accustomed to guerilla military operations who were required to adjust to conventional military operations and organizational culture.[81]

The South African Ministry of Defence issued its first defense white paper in May 1996, which was followed by a 1998 white paper and a December 1999 white paper on defense-related industries. The 1996 white paper stressed the transition from the apartheid government to a multi-racial democracy and mentioned that the new constitution established a framework for democratic civil-military relations in which civilian authorities retained control of the military. Key attributes of this document included the following:

- National security is sought to meet South Africa's political, economic, social, and cultural rights and needs while promoting and maintaining regional security.
- SANDF will adhere to international armed conflict law and to all international treaties it participates in.
- SANDF will have a primarily defensive orientation and posture.
- South Africa is committed to international arms control and disarmament and will participate in international efforts to contain and prevent small arms, conventional weapons, and weapons of mass destruction proliferation.
- Military force levels, weapons, and expenses will be determined by analysis of internal and external security environments and be subject to parliamentary approval.
- SANDF's primary role will be defending South Africa against external military aggression. Its deployment for internal policing will be limited to exceptional circumstances and require parliamentary approval and oversight.
- Defense policy and military activities will be transparent enough to ensure meaningful parliamentary and public scrutiny without endangering the lives of military personnel or jeopardizing military operations.[82]

This document went on to maintain that South Africa did not and will not have aggressive policies toward any state, that it is not confronted by an immediate conventional threat and does not anticipate external military threats in the next five years, that the vast majority of military conflicts occur within states, that fault lines between north and south countries have marginalized Africa in global political and economic matters, that the absence of a conventional military threat gives SANDF the opportunity to rationalize and redesign its capabilities, and that SANDF needs to maintain a core defense capability due to the unpredictability of potential future security requirements.[83]

The 1998 defense white paper reiterated many of these precepts while the 1999 defense industry white paper stressed that South Africa must have a defense industry capable of meeting its security requirements, which would remain under governmental operational control.[84] There have been no revisions of these white papers in the subsequent decade, but SANDF annual reports are good sources for noting evolutionary changes in South African military doctrine. The 1999/2000 SANDR *Annual Report* noted that conventional operations would consist of land operations that would be offensive, proactive, and reactive and intended to stop and destroy an enemy before it could enter South Africa, that air operations would focus on destroying hostile air forces on the ground, and that maritime operations would see hostile forces attacked and friendly shipping enhanced by defensive patrols and escorting. Nonconventional SANDF operations would focus on restoring law and order by supporting the South African Police Service, by conducting border control on land, sea, and air frontiers through high technology surveillance and rapid reaction forces, and by ensuring general area protection with high density and rapid reaction operations.[85]

There are several resources available for examining South African military doctrine and analysis of that doctrine. These include The Ministry of Defence (http://www.dod.mil.za/), whose Web site contents include annual reports from 2002–present, branches of South Africa's armed forces, including the Army's Web site (http://www.army.mil.za/), which features a 2006 issue of *Army Journal,* the Air Force's Web site (http://www.af.mil.za/), which includes *Ad Astra Magazine* (2004–present), and Navy's Web site (http://www.navy.mil.za/). Research organizations featuring analysis of South African military doctrine include the Institute of Security Studies (http://www.iss.co.za/), whose contents include a variety of reports and the scholarly journal *African Security Review* (1992–present), and the African Center for the Constructive Resolution of Disputes (ACCORD) (http://www.accord.org.za/).

South Korea

Recent South Korean military history has been shaped profoundly by the Korean War, which left the Korean peninsula divided at the 38th parallel. There has never been a peace treaty ending this conflict and both North and South Korea maintain heightened levels of military readiness on what is arguably the world's most contentious military frontier. South Korean military policy has been influenced by the government's transition from a dictatorship to a democracy and its remarkable economic growth; its alliance with the United States, which has undergone periodic challenges and strains in recent years as South Korea has sought to assert greater independence over its security policy; the controversy over how to deal with North Korea, as exemplified by the "Sunshine Policy" in which South Korea sought to improve relations with North Korea and provide increased economic assistance to that country's Stalinist regime; international concern over North Korea's nuclear weapons program and the best way to respond to that program; and by the need to structure the South Korean military to most effectively address early 21st-century challenges.[86]

South Korea has published recent defense white papers in 2000, 2004, and 2006. A 2005 Defense Reform plan published by the Ministry of National Defence seeks to improve the quality of the South Korean military while reducing its manpower from 690,000 to 500,000 by 2020 in response to a declining male birth rate and increasing the quantity of purchased weapons systems. In addition, this reform plan urges that South Korea move its military from a primarily conscript-based force to a more professional force. This plan envisions paying for these changes by increasing the defense budget 11.1 percent annually between 2005 and 2015, and 7 percent annually through 2020.[87]

The plan also stresses that an invasion from North Korea remains South Korea's preeminent national security priority, although it acknowledges that Russia, China, and Japan could potentially threaten South Korea given the proper circumstances. Requirements of a successful invasion of South Korea would include a ground force of one million or more along with supporting naval and air forces. North Korea's military equipment is antiquated and would be of limited effectiveness against contemporary South Korean and U.S. military technology, which the North Koreans may try to counter asymmetrically with weapons of mass destruction (WMD). Key requirements for responding to a North Korean attack would include:

- *Forward Defense.* Stopping the invasion's ground component in the forward area to preclude breakthrough possibilities.
- *Rear Area Defense.* Protecting South Korea behind the front lines and countering attacks by long-range artillery, missiles, aircraft, and special forces.
- *Target Strike.* Destroying North Korean military targets, including WMD, and limiting damage to opposing civilians.
- *Territorial Offensive.* Offensive operations to recover captured South Korean territory, find and destroy major hostile threats, and potentially remove enemy leadership.
- *Strategic Defense.* Use strategic weapons to deter enemy WMD use and destroy hostile forces and WMD use if deterrence fails.
- *Stability Operations.* Military efforts to secure captured territory and population and sufficient ground forces for successful stabilization, which could reach one million personnel.
- Incorporating credible responses to threats against sea and aerial communication attacks.[88]

South Korea's Ministry of National Defense (http://www.mnd.go.kr/) provides a number of information resources on Korean military policy and doctrine, although some of these documents use ePapyrus e-book reader, which is difficult to load and use on English language computers. The Korean National Defense University (http://www.kndu.ac.kr/eng/) has some English language content, including its Research Institute on National Security Affairs, which lists the names of many publications, though none are accessible in English. The Korea Institute for Defense Analyses (http://www.kida.re.kr/), which is a government-funded research institute specializing in defense, has articles from the scholarly *Korean Journal of Defense Analysis* (1999–present) and listings of other publications they produce. Presenting more of these resources in English would help further knowledge about South and North Korean military doctrinal literature.

The Institute of Foreign Affairs and National Security (http://www.ifans.go.kr/) is also a good resource for analysis of Korean national security policy issues.

Taiwan

The Taiwan Strait between Taiwan and the Republic of China is one of the world's most contentious waterways. On one side of this body of water is China, which is growing in international political, diplomatic, and military influence. The other side of the strait features the nation of Taiwan, which was founded by opponents of China's Communist government in 1949. Ensuing decades have seen Taiwan experience increasing economic prosperity, evolve from an authoritarian anti-communist government to a vibrant democracy, and develop significant military capabilities.

Taiwan has, however, struggled to achieve international diplomatic recognition as only a few countries have normal diplomatic relations with it due to intense Chinese political and diplomatic pressure to brand Taiwan as a "renegade province." Under the Taiwan Relations Act passed by the U.S. Congress in 1979, following the establishment of U.S. diplomatic relations with China, Taiwan was given implicit assurances of military support from the United States that it would receive American backing in the event of a Chinese military attempt to reunite the island with the mainland. The Taiwan Relations Act is not an official U.S. military alliance with Taiwan and there have been ebbs and flows in U.S.-Taiwanese relations since this act, with some administrations being more diplomatically or militarily supportive of Taiwan than others. However, all U.S. administrations have been acutely sensitive to how U.S. policy toward Taiwan would affect the increasingly important bilateral relationship between Beijing and Washington. A significant body of literature documents the challenges the United States faces in its relationships with China and Taiwan and Taiwanese national security problems and opportunities.[89]

Taiwan's military faces a number of daunting security threats from China. These include a buildup of Chinese ballistic missiles targeted at Taiwan that appear to emphasize being able to destroy opposing air and naval forces by targeting radar, naval, and air bases without needing to achieve air superiority. This was vividly demonstrated when China conducted a series of live fire missile exercises in the Taiwan Strait during 1995 and 1996. These exercises exposed deficiencies in U.S.-Taiwanese communication and military interoperability capabilities and resulted in the United States deploying two aircraft carrier groups into the area to demonstrate U.S. concern.[90]

These Taiwanese military challenges are exacerbated by misleading and restrictive U.S. military policies that, on one hand, encourage U.S. weapons sales to Taiwan, but, on the other hand, impose petty bureaucratic restraints on Taiwan. These restrictions, based on dubious fears of offending the Chinese, include not allowing the Taiwanese to purchase the most technologically advanced U.S. military equipment, requiring Taiwanese military personnel to wear civilian clothes when training in the United States, and rejecting Taiwanese requests for equipment to maintain major weapons systems. A 2001 U.S. Senate Foreign Relations Committee report strongly recommended that the United States allow Taiwan to purchase advanced military technology, end petty restrictions on visiting Taiwanese officials and military officers, end restrictions on U.S. military travel to Taiwan for training, establishing direct secure links between the U.S. and Taiwanese militaries, increase cooperation with Taiwan in areas such as intelligence and information warfare, and unequivocally state that the United States will defend Taiwan if it is attacked. Declining Taiwanese defense expenditures—which dropped from 22.8 percent of the government's budget in 1994 to 14.4 percent in 2002—indicate a cross-straits security balance turning in Beijing's favor, particularly considering China's growing defense budget.[91]

Taiwanese military policymaking is determined by several agencies. The President is the preeminent official in this process, with other actors being the Office of the President and Vice President, the Office of the Premier, the National Security Council, Ministry of Foreign Affairs, Ministry of National Defense, General Staff Headquarters, and the National Security Bureau. Some Taiwanese legislators may also be influential in formulating Taiwan's national security policies.[92]

The most authoritative statement of Taiwanese military doctrine and policy is provided by the biennial *National Defense Report,* the most recent edition of which was published in 2008 by the Ministry of National Defense. This report provides exhaustive analysis of the nature of the Chinese threat to Taiwanese national sovereignty, as the following passage demonstrates:

> The PRC military is actively modernizing its military to serve as a foundation for becoming a global power, and it uses its rapid economic growth to actively develop modernized military capabilities and to account for future regional warfare requirements. It is actively accelerating research and manufacture of joint operation command systems, enhancing joint firepower for large-scale battles, building formidable anti-sea and air defense capabilities over the Taiwan Strait, and purchasing platforms to build rapid projection capabilities in order to enhance its contingency capabilities and destructiveness. It also continues to accumulate attack capabilities that can execute precision strikes against Taiwan's political, military, and economic targets to sabotage Taiwan's command mechanism and economic order; additionally, its aircraft and ships continue to expand the radius of their activities, which is not only to gather intelligence about Taiwan's hydrology and airspace, but also to test Taiwan's naval and aerial response time, which serves as a reference for military actions against Taiwan.[93]

Taiwan's military doctrine, seeking to counter this threat, relies on the maxim, "Resolute Defense, Effective Deterrence," as its national defense modus operandi. "Effective Deterrence" involves constructing a defense force featuring sufficient deterrence capabilities that will convince adversaries to abandon military invasion due to uncertainty in achieving victory and risk of suffering unacceptable losses. "Resolute Defense" refers to the actions Taiwanese forces will take if "Effective Deterrence" fails to prevent hostile forces from conducting offensive invasions. "Resolute Defense" involves rapidly mobilizing reserve forces and converging national defense capabilities to repel hostile forces and execute debilitating counterstrikes against enemies by joint air, sea, and land forces.[94]

Taiwanese military doctrine also involves its armed forces working to enhance national missile defense capabilities, augmenting long-range precision strike capabilities, maintaining open waterway access to and from Taiwanese waters, acquiring long-range early warning radar surveillance capabilities, integrating C4ISR battlefield management systems, acquiring new generation jet fighters to ensure air superiority, and augmenting information warfare capabilities.[95]

Useful sources for examining Taiwanese military doctrine and strategy as well as assessments of these subjects include the Ministry of National Defense (http://www.mnd.gov.tw/), the National Security Bureau (http://www.nsb.gov.tw/), although most of its information is in Chinese, National Defense University (http://www.ndu.edu.tw/), and Taiwan Security Research (http://taiwansecurity.org/).

United Kingdom

The United Kingdom is another important producer of historically significant and relevant contemporary military doctrine literature. Recent centuries have seen British military forces take part in a wide variety of global conflict zones. In recent years, British military forces have engaged in combat operations in areas as diverse as Afghanistan, Bosnia, Iraq, and Sierra Leone to support national security interests, collaborate with the United States as part of the close defense cooperation between these countries, and participate in NATO- or United Nations-authorized operations. These campaigns have produced significant quantities of literature by British and other sources documenting British military activities and creating doctrine for conducting and evaluating military activities covering land, maritime, aerial, and counterinsurgency operations.[96]

Development of a formal written corpus of British military doctrine has been a relatively recent historical phenomenon. The following passage from an analysis of British military doctrinal development reflects how a disdain for written military doctrine within British military culture has shifted to an appreciation for its value because of major changes in Britain's security environment created by the collapse of the Soviet Union and the emergence of a multi-polar world.

> Traditionally, British officers did not care about intellectual debate and felt deep reluctance towards any formal writings. At best, some sort of doctrine existed as tactical instruction manuals. However, they were considered to be something for the classroom but irrelevant in the field. Operational experience was handed down informally, often by word of mouth, through generations of officers. It remained compartmentalized within the military's various groupings. In the absence of formal statements on the overall role of the British Armed Forces, a common starting point for the study of conflict did not exist. In such an organizational culture, innovation was left to coincidence, largely steered by what was already known or physically available.[97]

This eventually produced a transformation in British attitudes toward written military doctrine as reflected in the following observation:

> The period after 1989 witnessed the reversal of this attitude. A British 'soldier-scholar' emerged who was interested in the conceptual development of his

> institution. Formal doctrine statements started to be published, with particular efforts on the military-strategic level. This process intensified and by the end of the decade doctrine was firmly embedded within Britain's armed forces, giving evidence of an institution in search of more coherence in its conceptual bedrock.[98]

The traditional British reticence, if not antagonism, toward developing a coherent written corpus of military doctrine dissipated in the emergence of the post–Cold War world. British military leaders recognized that the multiplicity of security options in a multi-polar international security environment that could involve the use of their military required the development of a theoretical doctrinal framework. Such a framework would need to justify putting British forces into activities such as peacekeeping, urban warfare, and counterterrorism, which conventional national military forces had not seen as falling into their areas of responsibility.

Examples of this enhanced British military conceptual coherence began to be reflected in the publication of numerous individual military branch doctrinal publications and joint national strategy documents, reflecting contributions by the Ministry of Defence (MOD), individual armed services branches, and the policies of Prime Ministers Margaret Thatcher, John Major, Tony Blair, and Gordon Brown. 1989 saw the publication of *Design for Military Operations: The British Military Doctrine,* which presented the British Army's maneuver warfare approach, July 1990 saw the government present to the House of Commons its *Options for Change* statement, which called for reducing the number of British military personnel, and July 1991 saw the Royal Air Force (RAF) publish *Air Power Doctrine AP 3000* as the RAF's first high-level doctrine document since 1957.[99]

April 1993 saw the Army establish an Inspectorate General of Doctrine and Training, which would be reorganized as the Directorate General of Development and Doctrine (DGDD) the following year. January 1995 saw the Army publish *Wider Peacekeeping,* which was that service's first post–Cold War attempt to formulate peacekeeping doctrine. November 1995 saw the Royal Navy issue *The Fundamentals of British Maritime Doctrine* as its statement on post–Cold War maritime power. An updated version of *Design for Military Operations* was issued in January 1996, and the following January saw the establishment of the Joint Services Command and Staff College (JSCSC) to stress the importance of training and educating officers to conduct joint military operations. January 1997 also saw publication of *British Defence Doctrine JWP 0–01,* which was the first example of a joint doctrine document produced by the British military.[100]

The May 1997 election of Tony Blair's Labour Government would prompt the publication of further British military strategic and doctrinal documents. July 1998 produced this government's *Strategic Defence Review: Modern Forces for the Modern World,* which attempted to consolidate post–Cold War security and defense policy. This document emphasized that there were a wide range of threats to national security, including ethnic and religious conflict, population and environmental pressures, competition for scarce resources, drugs, terrorism, and crime. Such threats were described as smaller than Cold War threats, but they were operationally demanding on the forces engaged to address them, as reflected by events in areas such as Northern Ireland and Bosnia.[101]

October 2000 saw the establishment of the Joint Doctrine and Concepts Centre (JDCC) as the first organization for joint doctrine development. The 9/11 terrorist attacks also produced changes in British military doctrine and policy, as British forces assisted U.S. antiterrorist military operations in Afghanistan and Iraq, with the latter operations being particularly controversial within some sectors of British public opinion. July 2002 saw publication of *Strategic Defence Review—A New Chapter,* which sought to describe Britain's response to terrorism in the emerging post–9/11 security environment. Key points stressed in this document were calling for real increased defense spending by 1.2 percent per year over the next three years, the danger of imposing excessive burdens on military forces through repeated deployments, the need to develop extra strategic lift and communications capabilities for operations beyond counterterrorism, and examining how overall strategic priorities might provide additional emphasis to developing rapid reaction forces.[102]

December 2006 saw the British government address the continuing relevance of its nuclear deterrent during a time that emphasized the preeminence of counterterrorism in military operations. This document stressed that the United Kingdom could reduce its stockpile of operationally available warheads to less than 160, which was 20 percent below the number specified in the 1998 *Strategic Defence Review;* that conditions for complete UK nuclear disarmament do not exist due to the lack of progress in reducing nuclear stockpiles and the absence of global adherence to not proliferate nuclear weapons; and that there was a need to retain a nuclear deterrent to support collective NATO security in the Euro-Atlantic area. It went on to mention that a nuclear deterrent is necessary for informing adversaries that the cost of an attack against UK vital interests may result in nuclear retaliation against them; that the number of nuclear weapons-armed states may increase in subsequent decades; and that Britain needs to maintain ambiguity about when and if it might need to use its nuclear deterrent because it cannot tell an adversary what it would not do to defend vital national interests.[103]

March 2008 saw publication of Britain's most current national security strategy publication, *The National Security of the United Kingdom: Security in an Interdependent World.* This document, produced by the British Cabinet Office and the new premiership of Gordon Brown, mentions that while no state directly affects British national security, there are interconnected threats and risks to its security from international terrorism, weapons of mass destruction, conflicts involving failed states, pandemics, and transnational crime. Factors that can exacerbate these threats and risks also include climate change, energy competition, poverty, poor governance, demographic changes, and globalization.[104]

This document goes on to describe that general characteristics of Britain's response to these security threats would include:

- Grounding national security policy in core values such as human rights, the rule of law, legitimate and accountable government, justice, freedom, tolerance, and equal opportunity.
- Being hard-headed about risks, national aims, and capabilities to respond to these threats by national assets and international allies.
- Tackling security challenges early, if possible, with emphasis on preventive action capabilities.
- Favoring a multilateral approach to these problems, with emphasis on collective action through the United Nations, European Union, and NATO.
- Favoring a domestic partnership approach involving collaboration between the military, intelligence agencies, police, and border security personnel.
- Developing a more integrated approach in government policymaking that recognizes that distinctions between domestic and foreign policy are not helpful in a globalized world.
- Retaining strong, balanced, and flexible capabilities to better predict future threats, while recognizing that surprises will occur.
- Continuing to invest, learn, and improve ways to strengthen national security, while monitoring the effects of policies and actions to learn from experience.[105]

This document also reiterated the British Government's commitment to maintaining strong conventional forces with the ability to deter and respond to various state-led threats. It also asserts that British military spending will emphasize force capability over quantity and that defense procurement will stress supporting current operations and enhancing capabilities in areas such as strategic airlift, support helicopters, protected patrol vehicles, and surveillance and personal equipment.[106]

MOD's Development Concepts and Doctrine Center (http://www.mod.uk/DefenceInternet/MicroSite/DCDC/) is a key resource for British joint military doctrine documentation. Examples of British joint military doctrine publications here include *Joint Air Operations: Interim Joint Warfare Publication 3–30* (2003), *Joint Operations Planning 5–00* (2004), *Logistics for Joint Operations 4–00* (2007), and *DCDC Global Strategic Trends Program 2007–2036* (2007).

Additional sources of British military doctrine information as well as analysis of that doctrine include the Ministry of Defence (http://www.mod.uk/), the Defence Academy (http://www.da.mod.uk/), and its component entities, such as the Joint Services Command and Staff College, Royal College of Defence Studies, and Advanced Research and Assessment Group. Additional British military resources for military doctrine analysis include the Royal Air Force (RAF) (http://www.raf.mod.uk/), whose Web site features the 1999 edition of *AP3000 British Air Power Doctrine* and issues of the scholarly journal *Air Power Review* (2000–present), which is published by the RAF's Centre for Airpower Studies (http://www.airpowerstudies.co.uk/), the British Army (http://www.army.mod.uk/), the Royal Marines (http://www.royalmarines.mod.uk/), and the Royal Navy (http://www.royal-navy.mod.uk/).

Nonmilitary British sources analyzing British military doctrine include the Centre for Defence and International Security Studies (http://www.cdiss.org/), King's College London Defence Studies Department (http://www.kcl.ac.uk/schools/sspp/defence/), Oxford Research Group (http://www.oxfordresearchgroup.org.uk/), and Royal United Services Institute for Defence and Security Studies (http:// www.rusi.org/).

This chapter has sought to illustrate the rich variety of global military doctrinal documents that are readily accessible to dedicated researchers. This literature reflects numerous perspectives on historic, contemporary, and emerging military doctrinal issues facing these countries as they confront various national security issues. Different cultural and political factors account for the military doctrines advocated by these countries throughout their histories. The following passage from an assessment of British military doctrine is also applicable to the importance of studying military doctrine in all countries due to the insights such doctrine can provide to the military policymaking of these countries:

> Doctrine is a dialogue between the past and present for the benefit of the future. To identify the right lessons requires a genuine interaction between doctrine, training, education and operational command. The ideal doctrine therefore combines well-proven experience with imaginative thinking. In this context, it is of paramount importance that the study of past operations is carried out carefully and as objectively as possible so that historical observations are not misused for merely justifying a specific line of thought. At the same time, doctrine must not be rigid but allow sufficient room for flexibility and adaptation, since each conflict brings distinct circumstances. Pragmatic solutions for current military problems and creativity for future scenarios can only flourish in the absence of rigidity. A military organization with such an open intellectual attitude is less likely to fall for the trap most often quoted by historians—the observation that the military usually prepares for the last war instead of the next one.[107]

Notes

1. John A. Cope and Laurita Denny, *Defense White Papers in the Americas: A Comparative Analysis* (Washington, DC: National Defense University, Institute for National Strategic Studies, 2002), http://purl.access.gpo.gov/GPO/LPS82529 (accessed July 24, 2008).

2. For an overview, see Jeffrey Grey, *A Military History of Australia,* 3rd ed. (Port Melbourne, VIC: Cambridge University Press, 2008). For more detailed coverage, see Oxford University Press's Australian Centenary History of Defence series, including Jeffrey Grey, *The Australian Army* (South Melbourne: Oxford University Press, 2001); Alan Stephens, *The Royal Australian Air Force* (South Melbourne: Oxford University Press, 2001); David Stephens, ed., *The Royal Australian Navy* (South Melbourne: Oxford University Press, 2001); and David Murray, *Making the Australian Defence Force* (South Melbourne: Oxford University Press, 2001).

3. Australia, Department of Defence, *Defence 2000: Our Future Defence Force* (Canberra: Department of Defence, 2000), http://www.defence.gov.au/publications/wpaper2000.PDF (accessed July 25, 2008).

4. Australia, Army, *Fundamentals of Land Warfare LWD1* (Sydney: The Army, 2002), Chapter 1, p. 5 (unpaginated).

5. Seapower Centre Australia, "Organisation and Structure" (Canberra: Seapower Centre Australia, n.d.), http://www.navy.gov.au/spc/orgstrucmission.html (accessed July 25, 2008).

6. For recent assessments of Brazilian military policymaking, see Thomas E. Skidmore, *Brazil: Five Centuries of Change* (New York: Oxford University Press, 1999), 159–188; Craig Arceneaux, "Military Regimes and Transition Control in the Southern Cone and Brazil," *Journal of Political and Military Sociology* 29, no. 2 (2001): 259–274; Daniel Zirker, "Property Rights, Democratization, and Military Politics in Brazil," *Journal of Political and Military Sociology* 33, no. 1 (2005): 125–139; and Kai Michael Kenkel, "Language Matters: Security Discourse and Civil-Military Relations in Brazil," *Journal of Political and Military Sociology* 34, no. 2 (2006): 211–236.

7. Brazil, Ministry of Defense, *National Defense Policy 2005,* http://merln.ndu.edu/whitepapers/Brazil_English2005.doc (accessed July 28, 2008).

8. Ibid., 4–8.

9. Example surveys of Canadian military history include J. L. Granatstein, *Canada's Army: Waging War and Keeping the* Peace (Toronto: University of Toronto Press, 2002); Bernd Horn, ed., *The Canadian Way of War: Serving the National Interest* (Toronto: Dundurn Press, 2006); and Desmond Morton, *A Military History of Canada,* 5th ed. (Toronto: McLelland and Stewart, 2007).

10. See Sean M. Maloney, "Insights Into Canadian Peacekeeping Doctrine," *Military Review* 76, no. 2 (1996): 12–23; David Rudd, "Afghanistan, Darfur and the Great (Unexpected) Debate Over Canada's Military Role in the World," *Policy Options/Options Politiques* 27, no. 5 (2006): 53–57; Eric Wagner, "The Peaceable Kingdom?: The National Myth of Canadian Peacekeeping and the Cold War," *Canadian Military Journal* 7, no. 4 (2006): 45–54; David Pugliese, "Wakeup Call: Canadian Sovereignty, Economic Concerns Increase as Russia Flexes Muscle in the Arctic," *Seapower* 50, no. 10 (2007): 19–22; and Carl Ek and Ian Ferguson, et al., *Canada-U.S. Relations* (Washington, DC: Library of Congressional Research Service, 2008), 4.

11. Canada, Department of National Defence, *Canada First Defence Strategy* (Ottawa: Department of National Defence, 2008), 3, http://www.forces.gc.ca/site/focus/first/defstra_e.asp (accessed July 29, 2008).

12. Examples of this burgeoning literature include Alastair Ian Johnston, *Cultural Realism: Strategic Culture and Grand Strategy in Chinese History* (Princeton, NJ: Princeton University Press, 1995); Mark Burles and Abram N. Shulsky, *Patterns in China's Use of Force: Evidence From History and Doctrinal Writings* (Santa Monica, CA: Rand Corporation, 2000); James C. Mulvenon and Andrew N. D. Yang, eds., *Seeking Truth From Facts: A Retrospective on Chinese Military Studies in the Post-Mao Era* (Santa Monica, CA: Rand Corporation, 2001); Laurie Burkitt, Andrew Scobell, and Larry M. Wortzel, eds., *Lessons of History: The Chinese People's Liberation Army at 75* (Carlisle, PA: Strategic Studies Institute, U.S. Army War College, 2003); Ka-po Ng, *Interpreting China's Military Power: Doctrine Makes Readiness* (London and New York: Frank Cass, 2005); Andrew Scobell, "Is There a Chinese Way of War?: Review Essay," *Parameters* 35, no. 1 (2005): 118–122; and Roger Cliff, *Entering the Dragon's Lair: Chinese Antiaccess Strategies and Their Implications for the United States* (Santa Monica, CA: Rand Project Air Force, 2007).

13. U.S. Department of Defense, *Annual Report to Congress: Military Power of the People's Republic of China 2007* (Washington, DC: Department of Defense, 2007), 11–14.

14. *China's National Defense in 2006* (Beijing: Information Office of the State Council of the People's Republic of China, 2006), http://www.china.org.cn/english/features/book/194486.htm (accessed July 30, 2008).

15. *China's National Defense in 2006* (Beijing: Information Office of the State Council of the People's Republic of China, 2006), http://www.china.org/english/features/book/194485.htm (accessed July 30, 2008).

16. Examples of works on Estonia's history and security relationships include Toivo U. Raun, *Estonia and the Estonians,* 2nd ed. (Stanford, CA: Hoover Institution Press, 2001), 243–263; Pami Aalto, *Constructing Post-Soviet Geopolitics in Estonia* (London: Frank Cass, 2003); Steffen B. Rasmussen, "Estonian Security Perceptions in the Context of EU Enlargement: A Critical Discussion," *Baltic Defence Review* 11, no. 1 (2004): 154–173; and Toomas Riim, "Estonia and NATO: A Constructivist View on National Interest and Alliance Behavior," *Baltic Security and Defence Review* 8 (2006): 34–52.

17. See European Union, "Europa-European Countries-Estonia," (n.d.), http://europa.eu/abc/european_countries/eu_members/estonia/ (accessed July 31, 2008); and North Atlantic Treaty Organization, "NATO Press Releases," (2004), http://www.nato.int/docu/pr/2004/p04-047e.htm (accessed July 31, 2008) for information on when Estonia joined the EU and NATO. For an example of complications concerning relationships between the Baltic Republics and Russia, see Sven Gunnar Simonsen, "Compatriot Games: Explaining the 'Diaspora Linkage' in Russia's Military Withdrawal from the Baltic States," *Europe-Asia Studies* 53, no. 5 (2001): 771–791.

18. Estonia, *National Security Concept of the Republic of Estonia* (2004), 3–4, http://merln.ndu.edu/whitepapers/Estonia-2004.pdf (accessed July 31, 2008).

19. Ibid., 4–8.

20. Tomas Ries, *Cold Will: The Defence of Finland* (London: Brassey's Defence, 1988); Risto E. J. Penttila, *Finland's Search for Security Through Defence, 1944–89* (New York: St. Martin's Press, 1991); and David Kirby, *A Concise History of Finland* (New York: Cambridge University Press, 2006), 245–275.

21. Finland, Ministry of Foreign Affairs, "10 Years of EU Membership for Finland," (2008), http://virtual.finland.fi/netcomm/news/showarticle.asp?intNWSAID=25876 (accessed August 1, 2008).

22. See Gustav Hagglund, "Finnish Defence Policy Aims to Protect Against External Pressures," *NATO Review* 43, no. 4 (1995): 19–21 and David Arter, "Finland: From Neutrality to NATO?," *European Security* 5 (1996): 614–632.

23. Stephen J. Blank, *Finnish Security and European Security Policy* (Carlisle, PA: Strategic Studies Institute, U.S. Army War College, 1996), vii, 8, http://purl.access.gpo.gov/GPO/LPS12766 (accessed August 1, 2008).

24. Finland, Prime Minister's Office and Ministry of Defence, *Finnish Security and Defence Policy 2004: Government Report 6/2004* (Helsinki: Prime Minister's Office, 2004), 5–8, http://www.defmin.fi/files/311/2574_2160_English_White_paper_2004_1_.pdf (accessed August 1, 2008).

25. Ibid., 8–9.

26. Finland, Ministry of Defence, *Securely Into the Future: Ministry of Defence Strategy 2025* (Helsinki: Ministry of Defence, 2006), 5–11, http://www.defmin.fi/files/674/Securely_into_the_future_-_strategy_2025.pdf (accessed August 1, 2008).

27. Examples of literature examining historical influences on French military doctrine include Sten Ryning, "Shaping Military Doctrine in France: Decisionmakers Between

International Power and Domestic Interests," *Security Studies* 11, no. 2 (2001–2002): 85–115; Joseph Philippe Gregoire, *The Bases of French Peace Operations Doctrine: Problematical Scope of France's Military Engagements Within the NATO or UN Framework* (Carlisle Barracks, PA: U.S. Army War College, 2002); Tom Lansford, "Whither Lafayette?: French Military Policy and the American Campaign in Afghanistan," *European Security* 11, no. 3 (2002): 126–145; Gil Merom, *How Democracies Lose Small Wars: State, Society, and the Failures of France in Algeria, Israel in Lebanon, and the United States in Vietnam* (New York: Cambridge University Press, 2003); and Thomas D. Morgan, *The Fall of France and the Summer of 1940* (Arlington, VA: Institute of Land Warfare, Association of the United States Army, 2006).

28. France, President, *French White Paper on Defence and National Security* (Paris: President of France, 2008), 4, http://www.ambafrance-ca.org/IMG/pdf/Livre_blanc_Press_kit_english_version.pdf (accessed August 14, 2008).

29. Ibid., 4–8.

30. Ibid., 8–12.

31. France, Army, *Winning the Battle Building Peace: Land Forces in Present and Future Conflicts* (Paris: Centre de Doctrine d'Emploi Des Forces, 2007), 6–8.

32. Ibid., 12–13.

33. For a partial sampling of the immense monographic literature on this subject, see Arden Bucholz, *Hans Delbruck and the German Military Establishment: War Images in Conflict* (Iowa City: University of Iowa Press, 1985); Jehuda L. Wallach, *The Dogma of the Battle of Annihilation: The Theories of Clausewitz and Schlieffen and Their Impact on the Conduct of Two World Wars* (Westport, CT: Greenwood Press, 1986); James S. Corum, *The Roots of Blitzkrieg: Hans von Seeckt and German Military Reform* (Lawrence: University Press of Kansas, 1992); Antullio J. Echevarria II, *After Clausewitz: German Military Thinkers Before the Great War* (Lawrence: University Press of Kansas, 2000); Mary R. Habeck, *Storm of Steel: The Development of Armor Doctrine in Germany and the Soviet Union, 1919–1939* (Ithaca, NY: Cornell University Press, 2003); Robert M. Citino, *The German Way of War: From the Thirty Years' War to the Third Reich* (Lawrence: University Press of Kansas, 2005); and Jon Tetsuro Sumida, *Decoding Clausewitz: A New Approach to On War* (Lawrence: University Press of Kansas, 2008).

34. For critiques of the Schlieffen Plan, see Terence Zuber, *Inventing the Schlieffen Plan: German War Planning, 1871–1914* (New York: Oxford University Press, 2003); Terrence M. Holmes, "Classical Blitzkrieg: The Untimely Modernity of Schlieffen's Cannae Program," *Journal of Military History* 67, no. 3 (2003): 745–771; and Citino, *German Way*, 196–208.

35. For recent critiques of Blitzkrieg and its impact on German military doctrine, see Corum, *Roots of Blitzkrieg*; Robert M. Citino, *The Path to Blitzkrieg: Doctrine and Training in the German Army, 1920–1939* (Boulder, CO: Lynne Rienner, 1999); Alexandro B. Rossino, *Hitler Strikes Poland: Blitzkrieg, Ideology, and Atrocity* (Lawrence: University Press of Kansas, 2003); Habeck, *Storm of Steel*; Citino, *German Way*, 238–305; and Karl-Heinz Freiser and John T. Greenwood, *The Blitzkrieg Legend: The 1940 Campaign in the West* (Annapolis, MD: Naval Institute Press, 2005).

36. See Thomas-Durrell Young, *The 'Normalization' of the Federal Republic of Germany's Defense Structure* (Carlisle Barracks, PA: Strategic Studies Institute, U.S. Army War College, 1992); Jobst Schonfeld, *German-American Security Relations Within NATO and the UN* (Monterey, CA: Naval Postgraduate School, 1994); Thomas U. Berger, *Cultures of Antimilitarism: National Security in Germany and Japan* (Baltimore, MD: Johns Hopkins University Press, 1998); and Henning Tewes, *Germany, Civilian Power, and the New Europe: Enlarging NATO and the European Union* (Houndsmill, NY: Palgrave, 2002).

37. Philip Zelikow and Condoleezza Rice, *Germany Unified and Europe Transformed: A Study in Statecraft* (Cambridge, MA: Harvard University Press, 1995).

38. See Germany, *White Paper on the Security of the Federal Republic of Germany and the Situation and Future of the Bundeswehr* (Berlin: Defense Ministry, 1994), 2–3, http://www.resdal.org.ar/Archivo/d0000066.htm (accessed August 14, 2008); and "Last Russian Troops in Germany Head for Home," *Current Digest of the Post-Soviet Press* 46 (1994): 1–8.

39. See Mary Elise Sarotte, *German Military Reform and European Security* (Oxford: Oxford University Press, 2001); Martin Aguera, "Reforming the Bundeswehr: Defense Policy Choices for the Next German Administration," *Comparative Strategy* 21, no. 3 (2002): 179–202; Martin Kanz, "Dismissing the Draft: Germany Debates its Military Future," *Harvard International Review* 24, no. 4 (2003): 37–41; and Timo Noetzel and Benjamin Schreer, "Parliamentary Control of the Bundeswehr: The Need for Legislative Reform," *Stiftung Wissenschaft und Politik* (February 2007): 1–4.

40. Germany, *White Paper on the Security of the Federal Republic of Germany and the Situation and Future of the Bundeswehr*, 1–2.

41. Germany, *White Paper on the Security of the Federal Republic of Germany and the Situation and Future of the Bundeswehr.*

42. Germany, Bundesministerium der Verteidigung, *Defense Policy Guidelines* (Berlin: Bundesministerium der Verteidigung, 2003), 9–12, http://merln.ndu.edu/whitepapers/Germany_English2003.pdf (accessed August 14, 2008).

43. Sabine Collmer, *Information as a Key Resource: The Influence of RMA and Network-Centric Operations on the Transformation of the German Armed Forces* (Garmish-Partenkirchen: George C. Marshall Center for Security Studies, 2007), http://www.swp-berlin.org/en/common/get_document.php?asset_id=1800 (accessed August 14, 2008).

44. See Boris Wilke, "State-Building in Afghanistan?: Taking Stock of the International Presence in the Hindu Kush" (Berlin: Stiftung Wissenchaft und Politik, German Institute for International Security Affairs, 2004), http://www.swp-berlin.org/en/common/get_document.php?asset_id=1800 (accessed August 14, 2008); Michael Harsch, "Germany's Growing Afghan Dilemma" (Zurich: Center for Security Studies and Conflict Research, 2007), http://www.isn.ethz.ch/news/sw/details.cfm?ID=18423 (accessed August 14, 2008); and Timo Noetzel and Benjamin Schreer, "The German Army and Counterinsurgency in Afghanistan: The Need for Strategy" (Berlin: Stiftung Wissenschaft und Politik, German Institute for International Security Affairs, 2008), http://www.swp-berlin.org/en/common/get_document.php?asset_id=4752 (accessed August 14, 2008).

45. Germany, Federal Ministry of Defence, *White Paper 2006 on German Security Policy and the Future of the Bundeswehr* (Berlin: Federal Ministry of Defence, 2006), 5, http://merln.ndu.edu/whitepapers/Germany_White_Paper_2006.pdf (accessed August 14, 2008).

46. Ibid., 61, 63, 65, and 73.

47. For coverage of India's June 2008 announcement to create a military space capability, see Sudha Ranachandran, "India Goes to War in Space," *Asia Times Online*, June 18, 2008, http://www.atimes.com/atimes/South_Asia/JF18Df01.html (accessed August 19, 2008). For other analyses of Indian national security policymaking and military doctrine, see Amit Gupta, "Determining India's Force Structure and Military Doctrine: I WANT MY MIG," *Asian Survey* 35, no. 5 (1995): 441–458; Stephen Peter Rosen, *Society and Military Power: India and Its Armies* (Ithaca, NY: Cornell University Press, 1996); Brahma Chellaney, "After the Tests: India's Options," *Survival* 40, no. 4 (1998–99): 93–111; Swaran Singh, "Indian Debate on Limited War Doctrine," *Strategic Analysis* 23 (2000): 2179–2185; George Perkovich, *India's Nuclear Bomb: The Impact on Global Proliferation* (Berkeley: University of California Press, 2000); P. K. Chakravorty, "Artillery Revolution: An Indian Perspective," *Military Technology* 7 (2004): 82; Lowell Dittmer, ed., *South Asia's Nuclear Security*

Dilemma: India, Pakistan, and China (Armonk, NY: M.E. Sharpe, 2005); and Walter C. Ladwig III, "A Cold Start for Hot Wars?: The Indian Army's New Limited War Doctrine," *International Security* 32, no. 3 (2007/08): 158–190.

48. India, Ministry of Defence, *Annual Report 2007–08* (New Delhi: Ministry of Defence, 2008), 2, http://mod.nic.in/reports/AR-eng-2008.pdf (accessed August 19, 2008).

49. Ibid., 3–6.

50. Embassy of India, Washington, DC, "Draft Report of National Security Advisory Board on Indian Nuclear Doctrine" (Washington, DC: Embassy of India, 1999), 1, 3, http://www.indianembassy.org/policy/CTBT/nuclear_doctrine_aug_17_1999.html (accessed August 20, 2008).

51. Harsh V. Pant, "India's Nuclear Doctrine and Command Structure: Implications for India and the World," *Comparative Strategy* 24, no. 3 (2005): 278–279.

52. Arzan Tarapore, "The New Army Doctrine in Limited War," (New Delhi: Institute of Peace & Conflict Studies Military & Defence, 2004), 1–2.

53. For coverage of recent Indonesian history and the role played by the military in Indonesian government and policymaking, see "Defending Indonesia, Fifty Years on," *Asian Defence Journal* 10 (1995): 4; Robert Lowry, *The Armed Forces of Indonesia* (St. Leonards, AU: Allen & Unwin, 1996); M. C. Ricklefs, *A History of Modern Indonesia Since C. 1200* (Stanford, CA: Stanford University Press, 2001); Angel Rabasa and John Haseman, *The Military and Democracy in Indonesia: Challenges, Politics, and Power* (Santa Monica, CA: Rand Corporation, 2002); John Haseman, "Indonesia's Changing Role in the War on Terrorism," *Jane's Intelligence Review* 14, no. 11 (2002): 46–49; and Eric Heginbotham, "The Fall and Rise of Navies in East Asia: Military Organizations, Domestic Politics, and Grand Strategy," *International Security* 27, no. 2 (2002): 115–121.

54. Rabasa and Haseman, *Military and Democracy*, 25–26.

55. Ibid., 26–27.

56. Library of Congress, Federal Research Division, *Country Profile: Indonesia* (2004), 19, http://lcweb2.loc.gov/frd/cs/profiles/Indonesia.pdf (accessed August 20, 2008).

57. Ibid., 20–21.

58. Indonesia, Ministry of Defence, *Defending the County in the 21st Century* (Jakarta: Ministry of Defence, 2003), xi–xii, http://merln.ndu.edu/whitepapers/IndonesiaWhitePaper.pdf (accessed August 19, 2008).

59. Ibid., viii, ix.

60. For some of the literature on this topic, see Ariel Levite, *Offense and Defense in Israeli Military Doctrine* (Boulder, CO: Westview Press, 1990); Reuven Pedatzur, "Updating Israel's Military Doctrine," *IDF Journal* 22 (1991): 32–35; Frank K. Sobchak, "'Ah Harey'—Follow Me—Origins of the Israeli Junior Leadership Doctrine," *Military Intelligence* 19, no. 4 (1993): 20–23; Stuart A. Cohen, "The Israeli Defense Forces (IDF): From a 'People's Army' to a 'Professional Military'—Causes and Implications," *Armed Forces and Society* 21, no. 2 (1995): 237–254; Uri Bar-Joseph, ed., *Israel's National Security Toward the 21st Century* (Portland, OR: Frank Cass, 2001); Gabriel Ben-Dor, Ami Pedatzur, and Badi Hasisi, "Israel's National Security Doctrine Under Strain: The Crisis of the Reserve Army," *Armed Forces & Society* 28, no. 2 (2002): 233–255; Merom, *How Democracies;* Gregory R. Copley, "Israeli Strategic Doctrine: New Realities, New Responses," *Defense & Foreign Affairs Strategic Policy* 32, no. 11–12 (2004): 6–10; Martin Van Creveld, *Defending Israel: A Controversial Plan Toward Peace* (New York: Thomas Dunne Books/St. Martin's Press, 2004); Sergio Catignani, "Israel Defence Forces Organizational Changes in an Era of Budgetary Cutbacks," *RUSI Journal* 149, no. 5 (2004): 72–76; Uri Bar-Joseph, "The Paradox

of Israeli Power," *Survival* 46, no. 4 (2004–05): 137–156; Zeev Moaz, *Defending the Holy Land: A Critical Analysis of Israel's Security and Foreign Policy* (Ann Arbor: University of Michigan Press, 2006); Shlomo Brom, *From Rejection to Acceptance: Israeli National Security and Palestinian Statehood* (Washington, DC: United States Institute of Peace, 2007); and Efraim Inbar, *Israel's National Security: Issues and Challenges Since the Yom Kippur* (London: Routledge, 2008).

61. Bar-Joseph, "Paradox," 137–138.

62. Pedatzur, "Updating Israel's Military Doctrine," 32–35.

63. Sobchak, "'Ah Harey'," 20.

64. Copley, "Israeli Strategic Doctrine," 6–7; and Catignani, "Israeli Defence Forces," 72.

65. Examples of some of these analyses include Peter J. Vlakancic, *Marshal Tukhachevsky and the 'Deep Battle': An Analysis of Operational Level Soviet Tank and Mechanized Doctrine, 1935–1945* (Arlington, VA: Institute of Land Warfare, Association of the United States Army, 1992); Kimberley Marten Zisk, *Engaging the Enemy: Organization Theory and Soviet Military Innovation, 1955–1991* (Princeton, NJ: Princeton University Press, 1993); James H. Slagle, "New Russian Military Doctrine: Sign of the Times," *Parameters* 24, no. 1 (1994): 88–99; C. J. Dick, *Russia's 1999 Draft Military Doctrine* (Camberley, Surrey: Conflict Studies Research Centre, 1999); Aleksei Georgievich Arbatov, *The Transformation of Russian Military Doctrine: Lessons Learned From Kosovo and Chechnya* (Garmish-Partenkirchen, Germany: George C. Marshall European Center for Security Studies, 2000); Marcel De Haas, "An Analysis of Soviet, CIS, and Russian Military Doctrines 1990–2000," *Journal of Slavic Military Studies* 14, no. 4 (2001): 1–34; Christopher D. Jones, "Soviet Military Doctrine as Strategic Deception: An Offensive Military Strategy for Defense of the Socialist Fatherland," *Journal of Slavic Military Studies* 16, no. 3 (2003): 24–65; Habeck, *Storm of Steel;* Matthew Bouldin, "The Ivanov Doctrine and Military Reform: Reasserting Stability in Russia," *Journal of Slavic Military Studies* 17(2004): 619–641; Sergei Medvedev, *Rethinking the National Interest: Putin's Turn in Russian Foreign Policy* (Garmisch-Partenkirchen, Germany: George C. Marshall Center for Security Studies, 2004); Denis Trifinov, "Reversing Decline," *Jane's Defence Weekly* 42, no. 23 (2005): 27–29; Roger E. Kanet, ed., *Russia: Re-Emerging Great Power* (New York: Palgrave Macmillan, 2007); and Bradley A. Thayer and Thomas M. Skypek, "Russia Goes Ballistic," *The National Interest* 97 (2008): 61–68.

66. See Slagle, "New Russian," 88–90; and De Haas, "Analysis," 8–10.

67. See Arms Control Association, "Russia's Military Doctrine," (2000), 13, http://www.armscontrol.org/node/658/print (accessed September 2, 2008). See also De Haas, "Analysis," 21–32.

68. Arms Control Association, "Russia's Military Doctrine," 16–17.

69. Bouldin, "The Ivanov Doctrine and Military Reform," 624–627.

70. Trifinov, "Reversing Decline," 27.

71. Ibid.

72. See Trifinov, "Reversing Decline," 27 and Roy Allison, "Strategic Reassertion in Russia's Central Asia Policy," *International Affairs* 80, no. 2 (2004): 277–293.

73. Christopher Bayly and Tim Harper, *Forgotten Armies: The Fall of British Asia, 1941–1945* (Cambridge, MA: Harvard University Press, 2005), 106–155.

74. Resources describing Singapore's security policy and aspects of its military doctrine include Arujunan Narayanan, "Singapore's Strategy for National Survival," *Asian Defence Journal* 1 (1997): 6–7; Tim Huxley, *Defending the Lion City: The Armed Forces of Singapore* (St. Leonards, NSW: Allen & Unwin, 2000); Singapore, Ministry of Defence, *Defending Singapore in the 21st Century* (Singapore: Ministry of Defence, 2000); Felix K. Chang, "In

Defense of Singapore," *Orbis* 47, no. 1 (2003): 107–123; Pak Shun Ng, *From 'Poisonous Shrimp' to 'Porcupine': An Analysis of Singapore's Defence Posture Change in the Early 1980s* (Canberra: Australian National University, Strategic and Defence Studies Centre, 2005); "Singapore Homeland Security: The Official View," *Military Technology* 29, no. 3 (2005): 57–60; Yun Yun Teo, "Target Malacca Straits: Maritime Terrorism in Southeast Asia," *Studies in Conflict and Terrorism* 30 (2007): 541–561; and Robert Karniol and Tony Skinner, "Making the Connection: Country Briefing Singapore," *Jane's Defence Weekly* 44, no. 43 (2007): 22–27.

75. Teo, "Target Malacca Straits," 543.

76. Singapore, Ministry of National Defence, *The Fight Against Terror: Singapore's National Security Strategy* (Singapore: Ministry of National Defence, 2004), 11.

77. Ibid., 38–39.

78. Karniol and Skinner, "Making the Connection," 22–27.

79. Assessments of South African military history and doctrinal policy include Deon Fourie, "South Africa's Developing Security and Defence Policies," *RUSI Journal* 135, no. 2 (1990): 25–30; Robert J. Griffiths, "South African Civil-Military Relations in Transition: Issues and Influences," *Armed Forces & Society* 21, no. 3 (1995): 395–410; Garth Sheldon and Chris Alden, "Brave New World: The Transformation of the South African Military," *Comparative Strategy* 17 (1998): 345–362; Chris Bennett, "No Room for 'Nice to Haves'," *U.S. Naval Institute Proceedings* 126, no. 3 (2000): 44–47; M. Hough and L. Du Plessis, eds., *Selected Military Issues with Specific Reference to the Republic of South Africa* (Pretoria: Institute for Strategic Studies, University of Pretoria, 2001); Peter Liberman, "The Rise and Fall of the South African Bomb," *International Security* 26, no. 2 (2001): 45–86; "South Africa's New Defence Strategy," *Military Technology* 30, no. 1 (2006): 284–286; C. Homan, "Ambitious South African Armed Forces Struggle With Problems," *Militaire Spectator* 176, no. 5 (2007): 211–218.

80. Hough and Du Plessis, *Selected Military,* 7–10.

81. See Hough and Du Plessis, *Selected Military,* 11 and Sheldon and Alden, "Brave New World," 347–348.

82. South Africa, Ministry of Defence, *Defence in a Democracy: White Paper on National Defence for the Republic of South Africa* (Pretoria: South Africa Ministry of Defence, 1996), 4, 7–8.

83. Ibid., 16–17.

84. See South Africa, Ministry of Defence, *South African Defence Review* (Pretoria: Ministry of Defence, 1998); and South Africa, Ministry of Defence, *White Paper on the South African Defence Related Industries* (Pretoria: Ministry of Defence, 1999), 2.

85. Hough and Du Plessis, *Selected Military,* 31–32.

86. Literature examining the historical and contemporary evolution of South Korean military policy and doctrine includes Republic of Korea, Ministry of National Defense, *Defense White Paper 1992–1993* (Seoul: Ministry of National Defense, 1993); William J. Taylor et al., eds., *The Future of South Korean-U.S. Security Relations* (Boulder, CO: Westview Press, 1989); Young-Koo Cha and Kang Choi, "South Korea's Defense Posture," *Joint Force Quarterly* 7 (1995): 26–31; Roland Bleiker, *Divided Korea: Toward a Culture of Reconciliation* (Minneapolis: University of Minnesota Press, 1995); Yong-Pyo Hong, *State Security and Regime Security: President Syngman Rhee and the Insecurity Dilemma in South Korea, 1953–60* (New York: St. Martin's Press, 1999); Stephen Jin-Woo Kim, *Master of Manipulation: Syngman Rhee and the Seoul-Washington Alliance, 1953–1960* (Seoul: Yonsei University Press, 2001); Donald W. Boose Jr. et al., eds., *Recalibrating the U.S.-Republic of Korea Alliance*

(Carlisle Barracks, PA: U.S. Army War College Strategic Studies Institute, 2003); Bruce W. Bennett, *A Brief Analysis of the Republic of Korea's Defense Reform Plan* (Santa Monica, CA: Rand Corporation, 2006); U.S. Congress, House Committee on International Relations, *United States-Republic of Korea Alliance: An Alliance at Risk* (Washington, DC: Government Printing Office, 2006); Norman Friedman, "An Independent Role for South Korea?," *U.S. Naval Institute Proceedings* 132, no. 11 (2006): 90–91; and Jongryn Mo, "What Does South Korea Want," *Policy Review* 142 (2007): 43–55.

87. Bennett, *A Brief Analysis*, 1–2.

88. Ibid., 9–11.

89. Some of these works include Cheng Hsiao-Shih, *Party-Military Relations in the PRC and Taiwan: Paradoxes of Control* (Boulder, CO: Westview Press, 1990); Dennis Van Vranken Hickey, *The United States and Cross-Strait Rivalry: Strategic Partnership and Strategic Ambiguity* (Washington, DC: Atlantic Council of the United States, 1999); U.S. Congress, Senate Committee on Foreign Relations, *U.S. Defense Policy Toward Taiwan: In Need of an Overhaul* (Washington, DC: Government Printing Office, 2001); James M. Hughes, "China's Ballistic Missile Threat" *Journal of Social, Political, and Economic Studies* 27, no. 1 (2002): 3–22; Fang Hsu-hsiung, "The Transformation of U.S.-Taiwan Military Relations," *Orbis* 48, no. 3 (2004): 551–561; Martin Edmonds and Michael M. Tsai, eds., Taiwan's *Defense Reform* (New York: Routledge, 2006); Roger Cliff and David A. Shlapak, *U.S.-China Relations after Resolution of Taiwan's Status* (Santa Monica, CA: Rand Corporation, 2007); Michael D. Swaine et al., eds., *Assessing the Threat: The Chinese Military and Taiwan's Security* (Washington, DC: Carnegie Endowment for International Peace, 2007); Mumin Chen, "From Five No's to Referendum: The Making of National Security Policy in Taiwan," *Issues & Studies* 43, no. 3 (2007): 199–237; and Michael S. Chase, *Taiwan's Security Policy: External Threats and Domestic Politics* (Boulder, CO: Lynne Rienner, 2008).

90. Hughes, "China's Ballistic Missile Threat," 3; and Hsu-hsiung, "Transformation," 553–555.

91. See U.S. Congress, Senate Committee on Foreign Relations, *U.S. Defense Policy Toward Taiwan: In Need of an Overhaul,* 1, 10–11 and Hsu-hsiung, "Transformation," 558.

92. Chen, "From Five No's," 210–216.

93. Taiwan, Ministry of National Defense, *National Defense Report* (Taipei: Ministry of National Defense, 2008), 82.

94. Ibid., 134–135.

95. Ibid., 146, 164–166, 168–170, 172, and 179.

96. Partial samplings of this ample literature include Michael Codner, "Purple Prose and Purple Passion: The Joint Defence Centre," *RUSI Journal* 144, no. 1 (1999): 36–40; Alice Hills, *Doctrine, Criminality, and Future British Operations: A Half-Completed Understanding* (Camberley, Surrey: Strategic and Combat Studies Institute, 2000); A. A. Milton, "British Defence Doctrine and the British Approach to Military Operations," *RUSI Journal* 146, no. 6 (2001): 41–44; Julian Lindley-French, "Fighting Europe's Wars the British Way: The European Politics of Defence Doctrine," *RUSI Journal* 147, no. 2 (2002): 74–76; Markus Mader, *In Pursuit of Conceptual Excellence: The Evolution of British Military-Strategic Doctrine in the post–Cold War Era, 1989–2002* (Bern, Switzerland: Lang, 2004); Robert Fry, "Expeditionary Operations in the Modern Era," *RUSI Journal* 150, no. 6 (2005): 60–63; Jim Storr, "A Critique of Effects-Based Thinking," *RUSI Journal* 150, no. 6 (2005): 32–35; John Mackinley, "Is UK Doctrine Relevant to Global Insurgency?," *RUSI Journal* 152, no. 2 (2007): 34–38; Ken Young, "A Most Special Relationship: The Origins of Anglo-American Nuclear Strike Planning," *Journal of Cold War Studies* 9, no. 2 (2007): 5–31; and Andrew

Dorman, *Transforming to Effects-Based Operations: Lessons from the United Kingdom Experience* (Carlisle Barracks, PA: Strategic Studies Institute, U.S. Army War College, 2008).

97. Mader, *In Pursuit,* 22.

98. Ibid., 23.

99. Ibid., 358.

100. Ibid., 359–361.

101. Great Britain, Ministry of Defence, *Strategic Defence Review* (London: Ministry of Defense, 1998), 13–14.

102. See Mader, *In Pursuit,* 363–364; and Great Britain, Ministry of Defence, *Strategic Defence Review: A New Chapter* (London: Ministry of Defense, 2002), 28.

103. Great Britain, Ministry of Defence and Secretary of State for Foreign and Commonwealth Affairs, *The Future of the United Kingdom's Nuclear Deterrent* (London: Ministry of Defense, 2006), 8, 15, 18.

104. Great Britain, Cabinet Office, *The National Security Strategy of the United Kingdom: Security in an Interdependent World* (London: Cabinet Office, 2008), 3.

105. Ibid., 6–9.

106. Ibid., 45.

107. Mader, *In Pursuit,* 310.

CHAPTER 4

United Nations, North Atlantic Treaty Organization, and European Union Military Doctrine

National military doctrine documents are not the only sources students and scholars can use to study this topic. International government organizations (IGOs) such as the United Nations, North Atlantic Treaty Organization (NATO), and European Union have become increasingly involved in international military operations and have begun developing unique bodies of military doctrine as a basis for conducting such operations. This chapter will examine the origins and evolution of military operations conducted by these three IGOs and review sample doctrinal literature for these operations, which have generally focused on peacekeeping in various international locales.

United Nations

United Nations (UN) peacekeeping operations began in 1948 when the Security Council established an onsite operation of 36 unarmed military observers to preserve a truce after the first Arab-Israeli War.[1]

UN peacekeeping operations are established by the Security Council, which the United Nations charter designates as the organization primarily responsible for maintaining peace and security. However, financial aspects of peacekeeping operations are managed by the General Assembly. These organizations have delegated to the UN's Department of Peacekeeping Operations (DPKO) the responsibility for implementing UN peacekeeping objectives.[2]

Sixty-three UN peacekeeping operations have been conducted or were underway as of April 30, 2008, with 17 of these operations being active. These active operations involve 88,202 uniformed personnel, including soldiers, police, and military observers, from 117 countries. The financial cost of operations from July 1, 2007, to June 30, 2008, was approximately $6.8 billion, the cumulative financial cost of all operations from 1948 to the present is about $54 billion, and they have resulted in 2,468 peacekeeper fatalities, as of April 30, 2008.[3]

These operations have been assigned to a number of crisis areas around the globe and are denominated by a variety of acronyms. For instance, the United Nations Interim Administration Mission in Kosovo (UNMIK), established in June 1999, consists of 39 military observers, 1,917 police, 1,959 local civilians, and an overall personnel involvement of 4,503. Fifty-three fatalities have resulted from this mission, whose current annual budget is $210,676,800. Other examples of current UN peacekeeping missions included UN Organization Mission in the Democratic Republic of the Congo (MONUC), United Nations Mission in the Sudan (UNMIS), and United Nations Integrated Mission in Timor-Leste (UNMIT).[4]

The quality and effectiveness of UN Peacekeeping Operations is controversial. Supporters of these operations maintain that the UN is the most cost-effective means for grappling with international conflict and crises, that U.S. experiences in Afghanistan and Iraq mean that the United States cannot shoulder such operations on its own, that the United States should value the expertise UN members can bring to peacekeeping operations in diverse global environments, and that the UN, because of its perceived impartiality, can go into conflict areas where individual countries like the United States cannot. Critics of UN Peacekeeping operations assert that such operations give dangerous control to global authorities who may be antagonistic to U.S. national security interests, that the countries with the most capable militaries are less likely to contribute troops for peacekeeping, while those with the least capable militaries are the most likely to contribute their forces for such operations, that these forces are not given sufficiently liberal rules of engagement to effectively combat hostile operations against such missions, and that there are too many operational and cultural differences between members of these forces, who are trained in varying military traditions, to allow them to operate effectively together.[5]

An extensive corpus of military and political science literature exists on the performance and effectiveness of UN Peacekeeping Operations, reflecting a variety of perspectives. Topics addressed in this literature include whether the United States should participate in UN peacekeeping operations and whether U.S. forces should be commanded by foreign military leaders; managerial and financial support for such operations; the performance of UN peacekeepers in areas such as Bosnia, the Golan Heights, Haiti, and Sierra Leone; and the factors necessary for peacekeeping operations and subsequent conflict reconciliation to occur in these countries, including ethnic integration and incorporating combatants into the political process.[6]

UN peacekeeping doctrine documents are produced by DPKO and its Department of Field Support (DPS), and some are accessible through that organization's website (http://www.un.org/Depts/dpko/dpko/). These documents are divided into six major guidance series numbering 1000 to 6000. Documents in the 1000 series are known as Capstone Doctrine and cover the basic principles and critical concepts foundational to planning and conducting contemporary UN peacekeeping operations and the main factors affecting the success of those operations. Sample titles in this series include *United Nations Peacekeeping Operations: Principles and Guidelines* and *Handbook on United Nations Multi-dimensional Peacekeeping Operations.*[7]

The following excerpt from this document stresses that achieving a sustainable level of peace requires progress in at least four critical areas:

- Restoring the state's ability to provide security and maintain public order;
- Strengthening the rule of law and respect for human rights;
- Supporting the emergence of legitimate political institutions and participatory processes; and
- Promoting social and economic recovery and development, including safely returning or resettling internally displaced individuals and refugees uprooted due to this conflict.[8]

An additional section of this document describes UN Peacekeeping Operations as occurring in multiple, sometimes overlapping steps. These steps include the mission startup process, which involves pre-deployment where UN Headquarters negotiates Status of Mission and Status of Forces Agreements with the affected countries and parties; rapid deployment, where a small advance team arrives to begin establishing mission infrastructure and administrative systems; mission headquarters startup, which occurs when the mission leadership team arrives, command and control systems are established, and increasing numbers of support personnel arrive; and the establishment of substantive civilian, military, and police command capacities.[9]

Another noteworthy section of this UN doctrine document stresses the importance of maintaining local support for the mission. It warns that poor driving and vehicle accidents, along with poor waste management practices, can seriously degrade local support for mission legitimacy and popularity. This guidance also goes on to mention possible side effects to be aware of, including how staff conduct themselves socially; possible differences in what local societies may consider as gender-appropriate roles for women and mixed-gender working and socializing; how the economic impact of peacekeeping personnel may affect supplies and prices for housing, food, and other materials; and the importance of timely and effective public information activities to keep local populations informed about mission activities in order to retain their support.[10]

Documents in the 2000 series cover areas from headquarters support to operations and contain information on DPKO/DPS roles, responsibilities, and functions in supporting field missions. Examples of these documents are command, control, and executive direction; mission planning and budgeting; recruiting and force generation; deployment and mission initiation; political analysis and briefings; and reporting, monitoring, and operations management. Management and integration of operations in the field are covered in 3000 series documents. These documents seek to provide guidance on arrangements for effective planning, management, and integrating mission operational and support capabilities. Subjects addressed within this series include mission command and control; political analysis and diplomatic activity; mission planning; safety and security; crisis management; and peacekeeper conduct, welfare, and discipline.[11]

Information about multidimensional operations and guidance on employing military, police, and civilian capabilities within UN peacekeeping operation parameters are found in 4000 series documents, with such guidance being consistent with that provided by 1000 series publications. Topics covered within this series include political and civil affairs; military matters; police law enforcement; legal and judicial issues; correctional and prison matters; human rights; security sector reform; disarmament, demobilization, and reintegration of combatants into society; mine actions; and elections. Documents in the 5000 series feature guidance on integrating and supporting mission resources to meet mandate priorities in a timely and effective fashion. Examples of topics covered here include logistics support, movement control, strategic stockpile deployment, aviation, surface transport, engineering, communications and information technology, medical support, finances, and procurement and contract management.[12]

Finally, 6000 series documents cover headquarters management and administration and set out managerial and administrative procedures for DPKO/DFS as a UN Secretariat specialized, field-focused, and operational entity. Documents within this series cover planning, budgeting and oversight, human resources and travel, and writing and records.[13]

DPKO organizational activities are carried out under the leadership of Under Secretary-General for Peacekeeping Operations Jean-Marie Guehenno, who was appointed to this position by former UN Secretary General Kofi Annan on October 1, 2000.[14] Supporting offices within DPKO that provide additional insight into UN military doctrine policies and practices include the Office of Military Affairs (http://www.un.org/Depts/dpko/milad/), the Office of Rule of Law and Security Institutions (http://www.un.org/Depts/dpko/dpko/orolsi.shtml), a Policy, Evaluation, and Training Division, which features an Integrated Training Services section (http://www.un.org/depts/dpko/dpko/ITS.shtml) and a Best Practices Section (http://www.un.org/Depts/dpko/lessons/), which contains analytical reviews of UN peacekeeping publications, including *Engaging Civil Society in Peacekeeping: Strengthening Strategic Partnerships Between United Nations Peacekeeping Missions and Local Civil Society Organisations During Post-Conflict Transitions* (2007) and *HIV/AIDS Knowledge, Practice, and Attitude Survey: UN Uniformed Peacekeepers in Haiti* (2007).

The UN's Disarmament, Demobilization, and Reintegration Resource Center (http://www.unddr.org/) provides additional information resources on UN peacekeeping operations. Examples of the rich varieties of reports available here from the UN and other organizations include *Forgotten Fighters: Child Soldiers in Angola* (2003), *Taking RR to the People: National Information and Sensitization Campaign Field Report: Liberia DDRR Program* (2005), *Defense Reform and Conversion in Albania, Macedonia, and Croatia* (2006), and *Democratic Republic of Congo: Disarmament, Demobilization, and Reintegration (DDR) and the Reform of the Army* (2007).

Additional resources on UN Peacekeeping are provided by the research guide produced on this topic by the UN's Dag Hammarskjold Library at (http://www.un.org/Depts/dhl/resguide/specpk.htm). This guide provides information for conducting research on this topic using UN documents such as Security Council proceedings and resolutions, Secretary-General reports, correspondence between the Secretary-General and the Security Council President, the text of Security Council resolutions establishing peacekeeping operations, and General Assembly reports on funding and administering peacekeeping operations.

North Atlantic Treaty Organization

The North Atlantic Treaty Organization (NATO) began in the aftermath of World War II as victorious allied powers sought to develop security structures to prevent the occurrence of another global conflagration like World War II. The war's conclusion saw drastic reductions in U.S. troop strength from 3,100,000 in 1945 to 391,000 in 1946 and a reduction in British troop strength from 1,321,000 to 488,000 in the same period. Subsequent attempts between the victorious allied powers to produce peace treaties failed due to Soviet obstructionism and Soviet determination to create satellite ideological governments in Eastern Europe.[15]

This increasing tension between the western allied powers and the Soviets gradually led to increasing security collaboration between the United States and western European countries. On March 17, 1948, Belgium, France, Luxembourg, the Netherlands, and the United Kingdom signed the Treaty of Brussels, promising to establish a joint defensive system while enhancing their existing economic and cultural ties. This approximate time period also saw U.S. Secretary of State George Marshall (1880–1959) and U.S. Senators Arthur Vandenberg (1884–1951) and Tom Connally (1877–1963) begin discussions on North Atlantic security matters. Negotiations between the United States, Canada, and the Brussels Treaty participants began on July 6, 1948. These negotiations and subsequent developments led these powers to formally invite Denmark, Iceland, Italy, Norway, and Portugal to sign the pact, the contents of which were made public on March 18, 1949. On April 4, 1949, the North Atlantic Treaty was signed in Washington by the foreign ministers of these countries, and ratification of this agreement by national parliaments occurred within five months.[16] West Germany's May 5, 1955, incorporation into NATO was another sign that the alliance would grow in the future, further integrating NATO into the emerging postwar European security architecture.[17]

NATO was intended to be a defensively oriented alliance with its military focus primarily on the European continent, although it also encompassed its North American members, including the United States and Canada. The North Atlantic Treaty that established the alliance had 14 articles that allowed for future expansion, but the most important of these articles was Article 5, which stated that an armed attack against any NATO member was to be considered an attack against all members and that Article 51 of the United Nations Charter gave these countries the individual or collective right to defend themselves and North Atlantic security in any manner they considered necessary.[18]

During its subsequent six-decade history, NATO's civilian and military policymakers have sought to develop political and military doctrine to carry out NATO security objectives according to existing international security realities. A key realization for these policymakers was that it was not politically viable for NATO to match the size superiority of Soviet bloc conventional forces. This resulted in an emphasis on the development of a strong and credible nuclear deterrent as the means of deterring a possible Soviet invasion of Western Europe. An early example of this was the U.S. Army's attempt to integrate pentomic divisions into its organizational structure between 1954 and 1959. These pentomic divisions were to be small and highly-mobile and capable of conducting both conventional and nuclear operations, with the latter receiving primary emphasis.[19]

The pentomic division structure proved unworkable, but the emphasis on nuclear deterrence was initially ratified with the May 23, 1957, approval of Military Committee Document (MC) 14/2, which stressed that if a general war occurred, NATO should "ensure the ability to carry out an instant and devastating nuclear counteroffensive by all available means and develop the capability to absorb and survive the enemy's onslaught." Although this document provided latitude for NATO conventional forces to conduct operations, its preeminent emphasis on nuclear deterrence is unequivocal.[20]

This strategy would be updated by the flexible response doctrine enunciated in MC 14/3, issued on January 16, 1968. Influenced by the advocacy of U.S. Secretary of Defense Robert McNamara (1916–), MC 14/3 articulated a policy of direct defense, initially emphasizing conventional forces in which the alliance would seek to defeat aggression at the level at which the enemy chose to fight, placing the burden of escalation upon invading forces. MC 14/3 gave NATO the option of deliberately escalating to nuclear force, but controlling the scope and intensity of combat by increasing the aggressor's cost and increasing the imminence of a nuclear response. Such escalatory steps could include demonstrative uses of nuclear weapons and selective nuclear strikes on Soviet bloc interdiction targets. A critical component of this strategy is reflected in the following declaration:

> So long as forces committed to NATO and the external nuclear forces supporting the alliance are able to inflict catastrophic damage on Soviet nuclear society even after a surprise nuclear attack, it is unlikely that the Soviet Union will deliberately initiate a general war or any other aggression in the NATO area that involves a clear risk of escalation to nuclear war.[21]

Flexible response remained the cornerstone of official NATO strategic doctrine for the next two decades. However, there was criticism of its ambiguous nature and belief that it did not reflect changing European strategic conditions and public opinion during the 1970s and 1980s. These criticisms were voiced in a 1988 article in the U.S. Army War College professional journal *Parameters*. This article stated that strategic parity between the United States and Soviet Union had eroded the credibility of threats of deliberate escalation and detracted from NATO's ability to use nuclear threats to deter non-nuclear attacks and halt Soviet advances if deterrence failed. This article also maintained that U.S. and Soviet acquisition of a wide spectrum of theater and strategic nuclear forces undercuts the rationale NATO uses to justify the deliberate escalation portion of its flexible response strategy and that NATO would not gain a military advantage from introducing nuclear weapons into a Warsaw pact-initiated war.[22]

Concern over the effectiveness of NATO's nuclear deterrent caused alliance policymakers to examine ways of bolstering the effectiveness of its conventional forces. An example of this was the Follow-on Forces Attack (FOFA) concept approved by NATO's Defense Planning Committee in 1984. FOFA sought to build a NATO capability to hold leading divisions of a Warsaw Pact conventional forces assault by launching effective conventional force interdiction and destruction attacks against enemy follow-on forces before their logistical and combat support could be brought to the front lines.[23]

The collapse of the Soviet Union ended the Cold War in the early 1990s and made it unnecessary for NATO to seek to implement its military doctrine against the Soviet bloc invasion it had been designed to counter. However, the collapse of the post–World War II European security architecture posed new challenges for NATO. These would be first demonstrated when the collapse of Yugoslavia created vicious internecine ethnic conflict in the former Yugoslav republics of Bosnia-Herzegovina and Kosovo, which would eventually compel external intervention.

An early attempt by NATO to formulate how to respond to the new post–Cold War security environment was its November 1991 Strategic Concept document. Key points of this document included containing the consequences of potential civil and interstate conflicts in Eastern Europe and the former Soviet Union; collaborative defense against any aggression directed at alliance territory but not operations in areas beyond; and ensuring the territorial integrity of member states as a means of enhancing European peace and stability.[24]

The generally status quo nature of this document, reaffirming NATO's relative passivity toward offensive military operations, would not last for long. In mid–1992, NATO members began assuming peacekeeping responsibilities in Bosnia to enforce United Nations economic sanctions against Serbia as part of the United Nations Protection Force (UNPROFOR), with British and French troops playing important roles in this. NATO assisted UNPROFOR by providing close air support, monitoring the no-fly zone over Bosnia, and shooting down four Bosnian Serb aircraft on February 28, 1994, in the alliance's first use of deadly force.[25]

Another major example of NATO's increasingly assertive use of military force was the intervention of NATO forces against Serbia in 1999 to end Serbian violence in the former Yugoslav republic of Kosovo. Operation Allied Force, which lasted from March 23–June 10, 1999, was a NATO aerial campaign against the Serbian regime of Slobodan Milosevic (1941–2006) that succeeded in compelling the Serbians to withdraw their military forces from Kosovo. In the aftermath of this conflict, a NATO-led Kosovo force (KFOR) was established to provide security in this Serbian province until a decision was made on its final status.[26]

These conflicts provoked extensive debate within NATO and the international security community as to the doctrine that should be used for emerging forms of military conflict affecting NATO members in European and other locales. Considerable writing emerged in the late 1990s and beyond as to whether NATO should conduct military operations in areas outside its European stronghold; how to conduct military operations in theatres of operations outside Europe; and how to structure and command NATO forces if they are engaged in such operations. The increasing number of former Warsaw Pact countries admitted to NATO during the 1990s and 2000s posed additional complications for NATO planning and policymaking on these subjects and called into question the alliance's future viability.[27]

The 9/11 terrorist attacks against the United States further transformed NATO policy and doctrinal stances. The day after the attack, NATO invoked Article 5 of its charter for the first time, which requires members to come to the defense of each other when attacked.[28] A significant demonstration of NATO interest in enhancing its military capability was its November 21, 2002, Prague Summit decision to create a NATO Rapid Response Force (RRF). RRF was envisioned as consisting of technologically advanced, deployable, interoperable, and sustainable forces with land, air, and sea assets ready to move quickly at NATO Council determination.[29] NATO also sought to enhance its ability to make agile responses to military crises by replacing fixed mobile headquarters with nine Rapid Reaction Headquarters; inaugurating a program to deal with proliferating mass destruction weapons; and strengthening intelligence sharing to include European and U.S. homeland security.[30]

The biggest change these attacks prompted NATO to make was the decision to begin operations in Afghanistan after U.S.-led forces overthrew the Taliban regime in response to its support of the Al Qaida terrorist perpetrators of the 9/11 attacks. NATO operations in Afghanistan began when the alliance assumed command of the International Security Assistance Force (ISAF) in August 2003. ISAF represents NATO's first mission outside of the Euro-Atlantic region and its focus was initially restricted to Kabul. However, UN Security Council Resolution 1510, passed on October 13, 2003, enabled ISAF to support the Afghan Government throughout the entire country.[31]

ISAF's organizational structure consists of four components, including ISAF Headquarters, which is responsible for providing operational-level direction and planning to the Kabul Multinational Brigade, conducting operational assignments in its area of responsibility, and assisting the Afghan national government and nongovernmental organizations; the Kabul Multinational Brigade, which serves as ISAF's tactical headquarters responsible for planning, conducting, and patrolling civil-military operations on a daily basis; the Kabul International Airport, which assists Afghanistan's Ministry of Civil Aviation and Tourism in operating this airport; and the Provincial Reconstruction Teams (PRTs), which are civil-military partnerships responsible for providing security and reconstruction in Afghanistan's regions and helping the national government extend its authority over these regions. These responsibilities are executed by approximately 52,700 personnel from 36 NATO, nine partner, and two non-NATO/non-partner countries.[32]

Subsequent years have seen a resurgence of the Taliban campaign against the Afghan Government and ISAF forces. This insurgency has made some gains in its efforts to regain power in Afghanistan. Consequently, the quality of the ISAF/NATO response and campaigns in Afghanistan has undergone considerable scrutiny and criticism, and there is significant debate within the international security community regarding the likely success of ISAF and the overall quality of its Afghanistan operations. Some of this criticism and debate concerns whether individual ISAF country participants are committing enough troops to fight the Taliban and giving their forces sufficiently liberal rules of engagement to conduct effective combat operations. One example of this criticism is U.S. Secretary of Defense Robert Gates' January 2008 assertion that some NATO troops had not received proper counterinsurgency training. Although Gates' statement received critical response from other NATO countries, there is broad general agreement that the NATO/ISAF performance in Afghanistan will have a profound influence on NATO's political endurance and military operational viability in future international security crises.[33]

Developing an effective military doctrine, particularly for counterinsurgency operations in non-European combat theaters, will be critical if NATO is to be an organization capable of conducting successful military operations. Freeing itself from the compulsion to seek United Nations approval for its military actions will also be another demonstration that NATO is willing to serve as an effective force capable of conducting successful military operations.

Access to historical and contemporary NATO military doctrine resources is provided through a number of resources. These include the NATO E-Bookshop (http://193.219.98.16), the NATO Online Library (http://www.nato.int/docu/home.htm), NATO Archives (http://www.nato.int/archives/), NATO Standardization Agreements (http://www.nato.int/docu/standard.htm), and Parallel History Project on Cooperative Security (http://www.php.ish.ethz.ch/collections/). Additional useful NATO doctrine resources include those provided by the Joint Air Power Competence Centre (http://www.japcc.de/) in Kalkar, Germany, including *JAPCC Journal;* NATO's flagship periodical, *NATO Review* (http://www.nato.int/docu/review/), which is available online from January 1991–present; and publications produced by the NATO Defense College (http://www.ndc.nato.int/) in Rome.

European Union

The European Union began in the aftermath of World War II as states in Western Europe sought to work together to promote greater political and economic cooperation. French Foreign Minister Robert Schuman (1886–1963) presented a plan in 1950 to combine French and German coal and steel production into one organization and invited additional European countries to participate in this initiative. This would eventually result in Belgium, France, Italy, Luxembourg, the

Netherlands, and West Germany signing the Treaty of Paris to establish the *European Coal and Steel Community* (ECSC), which began operations in 1952.[34]

In 1955, ECSC Foreign Ministers began pursuing further economic cooperation opportunities, with negotiations leading to the signing of two treaties in Rome on March 25, 1957, and the setting up of the *European Economic Community* (EEC) and the *European Atomic Energy Community* (EURATOM). Further consolidation of EEC institutions occurred with the July 1, 1967, creation of a single executive body establishing "European Communities" (EC) as the term used to describe the mechanism for transnational European cooperation. In 1973, Denmark, Ireland, and the United Kingdom were admitted to the EC. Another major enhancement in European cooperation was the February 1986 signing of the *Single European Act,* which sought to bring foreign policy cooperation within the parameters of EC policymaking.[35]

EC interest in developing a unified security policy and military doctrine was subordinated to NATO during the Cold War era. This passive stance began to change following the end of the Cold War and the Soviet Union's collapse. The Maastricht Treaty of February 7, 1992, which came into force on November 1, 1993, sought to create greater European political integration as a result of German reunification and Communism's Eastern European collapse by establishing the European Union (EU) and developing a common foreign and security policy (CFSP).[36]

Proclaiming that the EU had a CFSP did not actually mean that a structure for integrating foreign and security policy actually existed, let alone that it possessed a coherent and viable doctrinal structure for implementing such policy cooperation.

The EU attempted to rectify this by proclaiming a European Security and Defense Policy (ESDP) at the Helsinki Summit on December 10–12, 1999. The Helsinki Summit called for the development of a European Expeditionary Force (EEF), which was envisioned as being used for humanitarian and rescue missions, peacekeeping, and the use of combat forces in crisis management operations such as peacemaking. EEF was expected to consist of 50,000–60,000 forces with an additional 140,000 troops for supporting extended operations. A 5,000-member police force was also called for to supplement this force by providing crisis management expertise.[37]

Additional rationales and desired capabilities for the EEF included lessening European dependence on the United States through the procurement of sufficient air and sealift capabilities, logistics, and Command, Control, Communications, Computers, Information, Surveillance, and Reconnaissance (C4ISR) and combat support to deploy this force within 60 days and sustain it for a year. A U.S. Army War College assessment of these objectives contended that achieving the EEF would require European states to reform or abolish conscription; restructure and modularize their forces to permit multinational formation; make significant investments in airlift capabilities such as the Airbus 400M to develop a European Air Transport Command; and enhance sealift and sea power capabilities, while also drastically increasing its precision attack and C4ISR capabilities if it wished to conduct joint operations with the United States.[38]

The 1999 Amsterdam Treaty gave the EU's CFSP five principal objectives, including:

- Safeguarding EU common values, fundamental interests, independence, and integrity in conformity with the United Nations Charter;
- Strengthening EU security;
- Preserving peace and strengthening international security according to United Nations Charter principles and related EU principles, including those applying to external borders;
- Promoting international cooperation; and
- Developing and consolidating democracy and the rule of law, respect for human rights, and fundamental freedoms.[39]

This pact also called for the EU to define principles and guidelines for conducting the CFSP, decide on common strategies for implementing these policies, and adapt joint actions and common positions. On December 14–15, 2001, the EU's European Council meeting in Laeken, Belgium adopted a declaration on the ESDP's operational capability, which provided official recognition that the EU was now capable of conducting some crisis management objectives.[40]

The EU had committed 20 combat brigades, 20 independent combat battalions, approximately 130 ships, and 500 fighter aircraft to its expeditionary force capabilities, although these only represented a small percentage of total potential EU force capabilities. This level of force commitment lagged behind U.S. military capabilities and caused former NATO Secretary-General Lord Robinson (1946–) to describe Europe as a "military pygmy."[41]

A significant EU effort to enhance its limited military capabilities was made by the December 12, 2003, release of its overall European Security Strategy. *A Secure Europe in a Better World* was the title of this strategic document, which sought to enunciate a coherent military strategy for the EU. It began by mentioning that European forces had been deployed to places as diverse as Afghanistan, the Democratic Republic of the Congo, and East Timor over the past decade; asserted that Europe should be ready to share responsibility for global security and the establishment of a better world; and emphasized that security is a precondition of development.[42]

A Secure Europe goes on to describe five key threat categories that it sees affecting European and global security. These include:

- *Terrorism.* Terrorism seeks to undermine societal openness and tolerance, uses electronic networks to carry out its aims, and uses European countries as targets and bases for such activities;
- *Proliferation of Weapons of Mass Destruction.* The EU acknowledges that proliferation of WMD has been partially reduced by international treaties and export control agreements, but warns of emerging dangers in the Middle East and scientific advances that could increase the potency of such weapons and provide advances in missile technology;

- *Regional Conflicts*. Conflicts such as those in Africa's Great Lakes Region or closer to Europe can directly or indirectly threaten national interests, destroy lives and physical and social infrastructures, and fuel the demand for weapons of mass destruction;
- *State Failure*. Whether caused by bad governance, corruption, abuse of power, weak institutions, or lack of accountability, state failure can corrode states from within, as demonstrated by Somalia, Liberia, and Afghanistan under the Taliban; and
- *Organized Crime*. This is a concern because Europe is a prime target for this activity, which can include drug trafficking, sex trade, illegal immigration, and weapons trafficking, which can have links with terrorism.[43]

This document goes on to maintain that the EU had sought to deal with such threats and threat scenarios by adopting a European Arrest Warrant, taking steps to fight terrorist financing, reaching a mutual legal assistance agreement with the United States, and intervening to deal with regional conflicts and restore failed states in areas such as the Balkans, Afghanistan, and the Democratic Republic of the Congo. It also contended that the EU would seek to transform its militaries into more flexible and mobile forces to contend with new threats, that it would increase defense spending if necessary, and that it would systematically use shared assets to reduce military duplication and increase military capabilities over the intermediate future.[44]

Despite the issuance of *A Secure Europe*, it is inaccurate to say that there is a coherent European military doctrine or an autonomous structure for European forces to conduct truly effective and independent military operations outside of NATO or U.S. auspices. One critical factor to consider is the diverse defense and security traditions among EU members. France and Great Britain have long histories of assertively taking unilateral military action. Germany, Italy, and Spain, due to their recent undemocratic and militarily aggressive pasts still face the historical baggage of their external military actions, which kept Germany from participating in UN peacekeeping operations until the 1990s. Other countries, such as Finland, Ireland, and Sweden, have developed considerable experience and expertise in UN peacekeeping operations and are reluctant to see the EU take a more militarily assertive role in international politics.[45]

Another assessment of European military capability stresses that the Europeans are probably incapable of catching up with U.S. efficiency in conducting large joint military operations at a fast pace. It emphasizes that these forces represent nearly 30 countries and have tremendous training, language, cultural, and equipment differences that make it nearly impossible to build a coherent force whose combat efficiency approximates that of the United States'. However, if the EU chooses to focus on high-intensity war fighting, it will come to depend more on the military structures of states that are willing and able to emphasize war fighting, which would reinforce the ESDP's intergovernmental nature and augment the strength of Europe's most militarily capable states, including Britain and France.[46]

Factors that could lead individual European countries or the EU has a whole to employ military force include the need of former colonial powers to use force in former colonies, as France and Belgium did in Zaire (Democratic Republic of the Congo) in 1993 and as Britain did to support a fragile government in Sierra Leone in 2000; the need to secure access to essential natural resources such as oil from other states; external political pressure for such intervention, such as the United States seeking military support from NATO allies; the threat posed by the spread of ballistic missiles and weapons of mass destruction; growing ethnic unrest in states bordering Europe, such as turmoil in the Balkans, the Middle East, and North Africa; and internal politics that may influence decisions to intervene in overseas humanitarian crises in places such as Darfur in Sudan.[47]

While the EU does not have a formal military doctrine like traditional nation states, it has established procedures for international crisis management. It is able to plan for policing and light peacekeeping operations on the military level and relies heavily on NATO or the assets of its largest members for conducting peace enforcement operations.[48]

This limited doctrinal guidance has made it possible for the EU to have undertaken approximately 20 missions through its ESDP. According to the European Foreign and Security Policy Institute, these missions employed as many as 7,000 personnel in Bosnia, but they have primarily focused on more limited objectives, such as preventing Macedonian civil unrest; reforming the Congolese Army and Georgian judicial system; training Afghan and Iraqi police forces; monitoring the Rafah crossing point in Gaza; and implementing a peace agreement in Aceh, Indonesia. Although EU governments have close to two million personnel in their armed forces and their collective defense spending was nearly $318 billion as of Spring 2008, they can barely deploy and sustain 100,000 soldiers globally.[49]

One scholar describes the status of EU military doctrine as follows:

> the search for an autonomous EU military doctrine cannot be fulfilled in the short term without challenging the dominance of NATO in European security or developing alternative models of European and international governance. This is why, in the context of the current diluting of the European integration process, the ongoing war on terrorism and the lack of citizens' political engagement at the European level, the EU's military doctrine will be autonomous only to the extent that a few key powers will allow it to be.[50]

Information and discussion about EU military doctrine can be found in a variety of sources even if there is no coherent, officially documented corpus of pertinent literature such as the United States' Joint Electronic Library. Such resources may be found in political science journals and databases that index articles from these journals; through the EU's Common Foreign and Security Policy website (http://ec.europa.eu/external_relations/cfsp/intro/), including through selected documents on this site such as *Small Arms and Light Weapons: The Response of the European Union* (2001) and *The European Union and India: A Strategic Partnership for the 21st Century* (2006); through publications produced by the EU's Institute for Security Studies (http://www.iss-eu.org/), including its Chaillot Papers and Occasional Papers monographic series; governmental and military policymaking debates; and through work produced by U.S. and European international political and security-oriented research institutes.

Notes

1. "United Nations Peacekeeping 2004–2005 Policy Debate Topic," *Congressional Digest* 83 (2004): 193.

2. United Nations Dag Hammarskjold Library, *United Nations Documentation: Research Guide* (New York: United Nations Dag Hammarskjold Library, 2008), 1, http://www.un.org/Depts/dhl/resguide/specpk.htm (accessed June 11, 2008).

3. United Nations, Department of Peacekeeping Operations, *Background Note: 30 April 2008,* 1, http://www.un.org/Depts/dpko/dpko/bnote010101.pdf (accessed June 11, 2008).

4. Ibid., 2.

5. See "Pro & Con: Is UN Peacekeeping an Effective Program, Deserving of U.S. Support?" *Congressional Digest* 83 (2004): 212–223 and Richard Connaughton, "Time to Clear the Doctrine Dilemma," *Jane's Defence Weekly* 21, no. 14 (1994): 19–20.

6. For a representative sampling of articles on these and related topics, see William H. Lewis and John O. B. Sewall, "United Nations Peacekeeping: Ends versus Means," *Joint Force Quarterly* 1 (1993): 48–57; Michael A. Collings, *United States Support for United Nations Peace Operations: Where Are We? Where Are We Going?* (Maxwell Air Force Base, AL: Air University Press, 1995), http://handle.dtic.mil/100.2/ADA328421 (accessed June 11, 2008); Brendan O'Shea, "The Future of UN Peacekeeping," *Studies in Conflict and Terrorism* 25, no. 2 (2002): 145–148; James Dobins, John G. McGinn, Keith Crane et al., *America's Role in Nation-Building: From Germany to Iraq* (Santa Monica, CA: Rand Corporation, 2003); Dan Lindley, "UNDOF: Operational Analysis and Lessons Learned," *Defense & Security Analysis* 20, no. 2 (2004): 153–164; James Dobbins, "The UN's Role in Nation-Building: From the Belgian Congo to Iraq," *Survival* 46, no. 4 (2004–05): 81–102; Nancy C. Roberts and Raymond Trevor Bradley, "Organizing for Peace Operations," *Public Management Review* 7, no. 1 (2005): 111–133; and Sven Gunnar Simonsen, "Building 'National' Armies—Building Nations?: Determinants of Success for Postintervention Integration Efforts," *Armed Forces & Society* 33, no. 4 (2007): 571–590.

7. United Nations. Department of Peacekeeping Operations. Department of Field Support, *United Nations Peacekeeping Operations: Principles and Guidelines* (New York: United Nations, 2008), 93.

8. Ibid., 25.

9. Ibid., 62–64.

10. Ibid., 81–83.

11. Ibid., 93.

12. Ibid.

13. Ibid.

14. United Nations, "Head of Department," http://www.un.org/Depts/dpko/dpko/info/page1.htm (accessed June 11, 2008).

15. Lord Ismay, *NATO the First Five Years, 1949–1954* (Paris?: North Atlantic Treaty Organization, 1954), 1–4, http://www.nato.int/archives/1st5years/chapters/1.htm (accessed June 23, 2008).

16. For additional historical background on NATO's origins, see Ismay, *NATO,* 7–10; Francis H. Heller and John R. Gillingham, *NATO: The Founding of the Atlantic Alliance*

and the Integration of Europe (New York: St. Martin's Press, 1992); Peter Duignan, *NATO: Its Past, Present, and Future* (Stanford, CA: Hoover Institution Press, 2000); and Gustav Schmidt, ed., *A History of NATO: The First Fifty Years,* 3 vols. (New York: Palgrave, 2001).

17. John A. Reed Jr., *Germany and NATO* (Washington, DC: National Defense University Press, 1987), 44–46.

18. North Atlantic Treaty Organization, "The North Atlantic Treaty" (1949), 1–2, http://www.nato.int/docu/basictxt/treaty.htm (accessed June 23, 2008).

19. David S. Yost, "The History of NATO Theater Nuclear Force Policy: Key Findings from the Sandia Conference," *Journal of Strategic Studies* 15, no. 2 (1992): 229–230.

20. Gregory W. Pedlow, ed., *NATO Strategy Documents 1949–1969* (Brussels: Supreme Headquarters Allied Powers Europe in Collaboration with NATO International Central Staff Archives, 1997), x, http://www.nato.int/archives/strategy.htm.

21. See David S. Yost, "NATO and the Anticipatory Use of Force," *International Affairs* 83, no. 1 (2007): 45–48; "Final Decision on MC 14/3: A Report By the Military Committee to the Defence Planning Committee on Overall Strategic Concept For the Defense of the North Atlantic Treaty Organization Area," in *NATO Strategy Documents, 1949–1969,* ed. Gregory W. Pedlow (Brussels: Supreme Headquarters Allied Powers Europe in Collaboration with NATO International Central Staff Archives, 1997), 356, 358, 360; and Wallace J. Thies, "On NATO Strategy: Escalation and the Nuclear Allergy," *Parameters* 18, no. 3 (1988): 19. For a critical appraisal of the concept of flexible response involving conventional and nuclear forces as applied to the Eisenhower and Kennedy Administrations' Berlin policy, see Kori N. Schake, "Case Against Flexible Response: Berlin Policy and Planning in the Eisenhower and Kennedy Administrations" (PhD diss., University of Maryland, 1996).

22. Thies, "On NATO," 22. See also Ivo H. Daalder, *The Nature and Practice of Flexible Response: NATO Strategy and Theater Nuclear Forces Since 1967* (New York: Columbia University Press, 1991).

23. See Yost, "History of NATO," 232; and U.S. Congress, Office of Technology Assessment, *New Technology for NATO: Implementing Follow-on Forces Attack* (Washington, DC: Government Printing Office, 1987). For a late 1980s assessment of NATO conventional force capabilities, see James M. Garrett, *The Tenuous Balance: Conventional Forces in Central Europe* (Boulder, CO: Westview Pres, 1989).

24. NATO Ministerial Communique, "The Alliance's New Strategic Concept" (1991), 1–15, http://www.nato.int/docu/comm/49-95/c911107a.htm (accessed June 23, 2008).

25. Yost, "NATO and the Anticipatory Use of Force," 50.

26. Ibid., 53. For assessments of the Kosovo war, see U.S. Department of Defense, *Report to Congress: Kosovo/Operation Allied Force After Action Report* (Washington, DC: Department of Defense, 2000), http://purl.access.gpo.gov/GPO/LPS16504 (accessed June 23, 2008); Ivo H. Daalder and Michael E. O'Hanlon, *Winning Ugly: NATO's War to Save Kosovo* (Washington, DC: Brookings Institution Press, 2000); Michael W. Lamb, Sr., *Operation Allied Force: Golden Nuggets for Future Campaigns* (Maxwell Air Force Base, AL: Air University Press, 2002); John Norris, *Collision Course: NATO, Russia, and Kosovo* (Westport, CT: Praeger, 2005); and Dag Henriksen, *NATO's Gamble: Combining Diplomacy and Airpower in the Kosovo Crisis, 1998–1999* (Annapolis, MD: Naval Institute Press, 2007).

27. Examples of this literature include Mike Wells, "Reaction Force Reshapes NATO Doctrine," *International Defense Review* 29 (1996): 73–76; Ove Bring, "After Kosovo: NATO Should Formulate a Doctrine on Humanitarian Intervention," *Journal of Legal Studies* 10 (1999–2000): 61–66; William E. Odom, "Making NATO Interventions Work: An American

Viewpoint," *Strategic Review* 28, no. 2 (2000): 13–18; S. Collins, "NATO and Strategic PSYOPS: Policy Pariah or Growth Industry," *Journal of Information Warfare* 1, no. 3 (2002): 72–78; and Nicholas Fiorenza, "Transforming NATO Air Power: New 'Competence Center' to Open," *Armed Forces Journal* 142, no. 5 (2004): 13–14. For examples of writing on NATO expansion and the operational implications of such expansion, see Edward B. Atkeson, "NATO Expansion: A Military Critique," *Army* 47, no. 11 (1997): 18–22; Joseph Lombardo, "NATO Expansion Saddled by Host of Economic, Military Variables," *National Defense* 82, no. 534 (1998): 37; "NATO Expansion: Full Speed Ahead—but to Where?," *Defense Monitor* 27, no. 2 (1998): 1–8; Ryan C. Hendrickson, "The Enlargement of NATO: The Theory and Politics of Alliance Expansion," *European Security* 8, no. 4 (1999): 84–99; Zoltan Barany, "NATO Expansion, Round Two: Making Matters Worse," *Security Studies* 11, no. 3 (2002): 123–157; and Thomas F. Lynch III, "NATO Unbound: Out-of-Area Operations in the Greater Middle East," *Orbis* 49, no. 1 (2005): 141–154.

28. North Atlantic Treaty Organization, "Statement by the North Atlantic Council 12 September 2001," http://www.nato.int/docu/pr/2001/p01-124e.html (accessed June 24, 2008).

29. Examples of accounts of the Taliban's demise in 2001 include Stephen D. Biddle, *Afghanistan and the Future of Warfare: Implications for Army and Defense Policy* (Carlisle Barracks, PA: Strategic Studies Institute, U.S. Army War College, 2002); Eric E. Theisen, *Ground-Aided Precision Strike: Heavy Bomber Activity in Operation Enduring Freedom* (Maxwell Air Force Base, AL: Air University Press, 2003), http://purl.access.gpo.gov/GPO/LPS40017 (accessed June 24, 2008); and Robert S. Tripp et al., *Supporting Air and Space Expeditionary Forces: Lessons from Operation Enduring Freedom* (Santa Monica, CA: Rand Corporation, 2004).

30. Giovanna Bono, "The EU's Military Doctrine: An Assessment," *International Peacekeeping* 11, no. 3 (2004): 449.

31. North Atlantic Treaty Organization, "NATO in Afghanistan: Factsheet," (2008), 1–2, http://www.nato.int/issues/afghanistan/040628-factsheet.htm (accessed June 24, 2008).

32. Ibid., 1–5. For ISAF personnel in Afghanistan, including national breakdowns as of June 10, 2008, see North Atlantic Treaty Organization, "International Security Assistance Force: ISAF Regional Commands & PRT Locations," (2008), 1–2, http://www.nato.int/docu/epub/pdf/isaf_placemat.pdf (accessed June 24, 2008).

33. Examples of the continually growing literature on this subject include William R. Hawkins, "What Not to Learn from Afghanistan," *Parameters* 32, no. 2 (2002): 24–32; Anthony Davis, "Afghan Security Deteriorates as Taliban Regroup," *Jane's Intelligence Review* 15, no. 5 (2003): 10–15; Howard G. Coombs and Rick Hillier, "Planning for Success: The Challenge of Applying Operational Art in Post–Conflict Afghanistan," *Canadian Military Journal* 6, no. 3 (2005): 5–14; Stephen D. Biddle, "Allies, Airpower, and Modern Warfare: The Afghan Model in Afghanistan and Iraq," *International Security* 30, no. 3 (2005–2006): 161–176; Orville F. Desjarlais Jr., "On the Road to Restoration: Bagram Provincial Reconstruction Team Helps Build Bridges, Roads and Schools," *Airman* 50, no. 4 (2006): 30–35; Cyrus Hodes and Mark Sedra, *The Search for Security in Post–Taliban Afghanistan* (Abingdon, UK: Routledge for the International Institute for Security Studies, 2007); Armed Forces Press Service News Articles, "Gates Says NATO Allies 'Committed' to Mission in Afghanistan," (2008), http://www.defenselink.mil/news/newsarticle.aspx?id=48688 (accessed June 24, 2008); Timo Noetzel and Benjamin Schreer, "The German Army and Counterinsurgency in Afghanistan: The Need for Strategy," (Berlin: Stiftung Wissenschaft und Politik, German Institute for International and Security Affairs, 2008), http://www.swp-berlin.org/en/common/get_document.php?asset_id=4752 (accessed June 24, 2008);

and Robin Shephard, "NATO Summit: Fears for the Future," *The World Today* 64, no. 4 (2008): 4–6.

34. Ian Thomson, *The Documentation of the European Communities: A Guide* (London: Mansell Publishing Ltd., 1989), 1. For additional historical background on the European Union's origins, see Trevor Salmon and Sir William Nicol, eds., *Building European Union: A Documentary History and Analysis* (Manchester and New York: Manchester University Press and St. Martin's Press, 1997); Michael Burgess, *Federalism and European Union: The Building of Europe, 1950–2000* (London: Routledge, 2000); Craig Parsons, *A Certain Idea of Europe* (Ithaca, NY: Cornell University Press, 2003); and Desmond Dinan, *Europe Recast: A History of European Union* (Boulder, CO: Lynne Rienner, 2004).

35. Thomson, *Documentation of European Communities,* 1.

36. European Union, "SCADPlus: Treaty of Maastricht on European Union," (2007), 1–2, http://europa.eu/scadplus/treaties/maastricht_en.htm (accessed July 15, 2008).

37. Andrew M. Dorman, *European Adaptation to Expeditionary Warfare: Implications for the U.S. Army* (Carlisle Barracks, PA: U.S. Army War College Strategic Studies Institute, 2002), v, 12. Additional institutional background and analysis on the structural organization of ESDP can be found in Michael Smith, "The Framing of European Foreign and Security Policy: Towards a Post–Modern Policy Framework?," *Journal of European Public Policy* 10, no. 4 (2003): 556–575; and Hylke Dijkstra, "The Council Secretariat's Role in the Common Foreign and Security Policy," *European Foreign Affairs Review* 13, no. 2 (2008): 149–166.

38. Ibid., vi.

39. European Union. External Relations. "Common Foreign and Security Policy Overview," (2002), 1–2, http://ec.europa.eu/external_relations/cfsp/intro/index.htm (accessed July 15, 2008).

40. Ibid.

41. Stale Ulriksen, "Requirements for Future European Military Strategies and Force Structures," *International Peacekeeping* 11, no. 3 (2004): 459, 457.

42. European Union, *A Secure Europe in a Better World: European Security Strategy* (2003), 1–2, http://consilium.europa.eu/uedocs/cmsUpload/78367.pdf (accessed July 15, 2008).

43. Ibid., 3–5.

44. Ibid., 6, 12.

45. Bono, "The EU's Military Doctrine," 448.

46. Ulriksen, "Future European Military Strategies," 463.

47. Dorman, *European Adaptation,* 4–6.

48. Bono, "The EU's Military Doctrine," 453.

49. Daniel Keohane, "The Strategic Rise of EU Defense Policy," *Issues* 25 (2008): 6.

50. Bono, "The EU's Military Doctrine," 453–454.

CHAPTER 5

Monographic Scholarly Literature

The scholarly monograph or book is another important venue for communicating academic research findings. This has been particularly true in the humanities and social sciences in the western world and still retains valid in the early years of the 21st century, even though ongoing technological information dissemination transformations are altering scholarly publishing in numerous ways, including how individuals outside the academic community view scholarly research.[1]

Military doctrine research has produced a significant and continually growing scholarly corpus representing disciplines as diverse as history, military science, political science, and even military sociology. Such research has been published by scholars from universities and public policy research institutions and by professional military officers from the United States and other countries. This chapter will examine and annotate representative samples of this research. It will not evaluate the intellectual or scholarly merits or demerits of these works and how their authors approach their topics, but will aspire to document the rich variety of work that has been produced and continues to be produced that analyzes historical, contemporary, and emerging military doctrines practiced by militaries and their national leaderships throughout the world.

Numerous academic publishers in the United States and elsewhere produce works on military doctrine and strategy. Examples of such publishers include the University Press of Kansas, Cornell University Press, Texas A&M University Press, Air University Press, Frank Cass, and many others. An effective way to search for books on military doctrine in library online catalogs is by using Library of Congress Subject Headings (LCSH) as search terms. Sample LCSH searches include "military doctrine," "national security," or a country's name and the phrase, "military policy" (e.g., "United States–Military Policy"). It is also possible to narrow LCSH searches by countries, geographic regions, chronological dates, or specific military forces (e.g., "Military Doctrine–Germany–History–20th-Century," "National Security–Indonesia," and "Australia Army History 1945–1965").

Entries will include standard bibliographic citations, listings of publishers' monographic series that the entries may be part of, the entries' International Standard Bibliographic Number (ISBN) to facilitate purchase or ordering through Interlibrary Loan services, and Web URLs if these resources are freely available on the Internet.

Adams, Thomas K. *The Army after Next: The First Postindustrial Army.* Westport, CT: Praeger Security International, 2006. ISBN: 978-0-275-98107-5.

This work examines how the U.S. Army and Department of Defense (DOD) have sought to create the capabilities needed to produce the technologically driven Revolution in Military Affairs (RMA) and examines how RMA and transformation paradigms have affected U.S. military operations in Afghanistan and Iraq. Adams is particularly critical of how technological aspirations have excessively influenced post–9/11 U.S. military operations despite the battlefield realities of these conflicts. He believes that doctrine should drive technology instead of the converse, that airpower is only a supportive element of successful military policy, that you should fight the war you are in rather than one based on ideologically driven constructs, that the ability of enemies to adapt cannot be changed by digitization, and that securing victory is almost as important as achieving it, with the cases of stability forces and psychological operations in Afghanistan and Iraq being particularly important demonstrations of this.

Blaker, James R. *Transforming Military Force: The Legacy of Arthur Cebrowski and Network Centric Warfare.* Westport, CT: Praeger Security International, 2007. ISBN: 978-0-275-99427-3.

Blaker discusses the influential military doctrinal thought propounded by Admiral Arthur Cebrowski (1942–2005), who served as the Director of the Defense Department's Office of Force Transformation between 2001 and 2005. Key tenets of Cebrowski's thinking were that humanity was naturally competitive but not naturally warlike; that the United States must prepare for possible armed conflict; that being militarily effective and moral requires moving from indiscriminate attrition warfare to more discriminate uses of force; that information technology provides the mechanism for more effective and moral uses of military force; that the U.S. military should migrate toward a network-centric design that would facilitate better information flows between units and confront opponents with overwhelming complexity; and that since the primary source of military power is shifting to globally available technology, the United States must accelerate how quickly it shifts to new force design to adapt to constant, rapid technological change if it wishes to retain its military dominance.

Celik, Murat. *Comparison of the British and Canadian CIMIC and the U.S. CMO Doctrines to the NATO CIMIC Doctrine.* Monterey, CA: U.S. Naval Postgraduate School, 2005. Also available online at http://handle.dtic.mil/100.2/ADA443057.

Celik intends for his work to enhance the ability of Turkish armed forces to develop a national doctrine for civil-military cooperation (CIMIC).

He contends CIMIC doctrine is critically important for peacekeeping, peace enforcement, and combat operations, and that military forces must move beyond acquiring and retaining territory to retain the support of populations in areas of combat operations. A key point of this work is that Turkey is in a geopolitical position to make major contributions to conflict stabilization in its adjacent geographic region, and he uses illustrations of NATO, British, Canadian, and U.S. CIMIC doctrines as applied to operations in Bosnia and Kosovo to bolster this contention.

Mulvenon, James, and David Finkelstein, eds. *China's Revolution in Doctrinal Affairs: Emerging Trends in the Operational Art of the Chinese People's Liberation Army.* Alexandria, VA: The CNA Corporation, 2005. Also available online at http://www.cna.org/documents/doctrinebook/pdf.

This compendium of essays examines changes occurring in Chinese military doctrine during the 1990s, with particular emphasis on how doctrinal changes may be reflected in operational planning. Overall themes include China showing increasing concern over Taiwan, increasing distrust of U.S. intentions toward China, concern over India's increasing regional ambitions, uncertainty over Japan's evolution in military and regional affairs, and competition with neighboring Southeast Asian nations for South China Sea natural resources.

Matters addressed in these essays include the emergence of joint operations in Chinese military doctrine, evolutions in Chinese military strategy from 1987 to 1999, trends and developments in Chinese nuclear force modernization and nuclear use doctrine, implementing People's Liberation Army (PLA) artillery doctrinal reforms, joint aerospace campaign strategy and doctrine, contradictions in PLA doctrine and Taiwan operational scenarios, and Chinese visions of possible military operations in space.

Citino, Robert M. *The Path to Blitzkrieg: Doctrine and Training in the German Army, 1920–1939.* Boulder, CO: Lynne Rienner Publishers, 1999. ISBN 1-5558-7714-1.

Citino shows how the German army rebuilt itself from defeat in World War I and how, thanks to the efforts of General Hans von Seeckt (1866–1936) and other generals, it was able to evade Versailles Treaty restrictions and rebuild itself to become a formidable fighting force at the onset of World War II. *The Path to Blitzkrieg* demonstrates how German war-fighting doctrine was comprehensively reformed and how it developed the capabilities necessary to become a military force capable of launching effective offensive military operations. A particularly salient point is how the German military began to make effective use of combined arms doctrine in which land forces sought to work with airpower to achieve optimal military effect and how a key Blitzkrieg component was increasing the tempo of war in order to keep opponents off-balance.

Citino, Robert M. *The German Way of War: From the Thirty Years' War to the Third Reich.* Modern War Studies. Lawrence: University Press of Kansas, 2005. ISBN: 0-7006-1410-9.

Treatise providing detailed historical analysis of German military doctrine and strategy during a three-century period. It begins by describing the origins of German military doctrinal thinking during the reign of Prussian ruler Frederick William I (1640–1688). Later chapters discuss the revolution in Prussian/Germanic military thought during the reign of Frederick the Great (1740–1786), with particular emphasis on the Seven Years War (1756–1763); defeats and recovery during the Napoleonic Wars; the impact of Karl von Clausewitz's *Vom Krige (On War)* on German and global military thinking; the role of General Helmuth von Moltke the Elder (1800–1891), who served as Prussian General Staff Chief from 1857 to 1888, in shaping German military policy to achieve national unification; and how these cumulative trends and policies produced military successes and crushing military failures during World Wars I and II.

Cliff, Roger, Mark Burles, Michael S. Chase, Derek Eaton, and Kevin L. Pollpeter. *Entering the Dragon's Lair: Chinese Antiaccess Strategies and Their Implications for the United States.* Santa Monica: Rand Project Air Force, 2007. ISBN: 0-8330-3995-4. Also available online at http://rand.org/pubs/monographs/MG524/.

This appraisal examines concerns that China might employ "antiaccess" strategies that would limit deployment of U.S. forces into combat theaters, restrict the locations from which U.S. forces could effectively operate, or compel opposing forces to operate from more remote combat locations than they would usually prefer. Developing such antiaccess strategies is particularly important for potential future U.S. military opponents given the tremendous technological and conventional military force superiority the United States is likely to enjoy in such confrontations.

Topics addressed in specific chapters include how Department of Defense publications such as the *Quadrennial Defense Review* have addressed the antiaccess challenge; attributes of contemporary Chinese military strategy, emphasizing local war under high technology conditions; Chinese military strategy components with possible implications for U.S. theater access, such as attacks on computer networks, satellites, sea lanes, ports, and aircraft carriers; examining the potential results of successful Chinese attacks against these assets; and possible ways for the United States to counter such antiaccess threats, including deploying air and missile defense systems near critical facilities, diversifying aircraft basing options, expanding counters to anti-satellite attacks, and enhancing early warning tactical and strategic capabilities.

Clodfelter, Mark. *The Limits of Air Power: The American Bombing of North Vietnam.* New York: The Free Press, 1989. ISBN: 0-02-905990-9.

This work seeks to evaluate the military effectiveness of the U.S. aerial bombing of North Vietnam between 1965 and 1972. Using Clausewitzian methodology, it focuses on the Rolling Thunder campaign from 1965 to 1968 and the Linebacker I and II operations of 1972 to examine how U.S. political objectives and military doctrine impacted U.S. bombing strategy. Clodfelter maintains that some U.S. goals restricted air power's application, such as preserving a non-Communist South Vietnam while limiting the use of military force to avoid direct Soviet or Chinese intervention. He also maintains that bombings' political effectiveness can be diminished by various military and operational limitations, such as doctrine, enemy defense, technology, geography, and weather. His ultimate conclusions are that Vietnam saw American policymakers counter a war that diverged from previous expectations, experience, and doctrine; that Johnson and his advisors failed to provide clear military airpower objectives, that U.S. military airpower objectives did not integrate with Johnson's political goals or insurgent warfare; and that aerial bombing doctrine is best equipped for a fast-paced conventional war, not guerilla warfare.

Corum, James S. *The Roots of Blitzkrieg: Hans von Seeckt and German Military Reform.* Modern War Studies. Lawrence: University Press of Kansas, 1992. ISBN: 0-7006-0541-X (Cloth); 0-7006-0628-9 (pbk).

Corum seeks to describe how the German military sought to rebuild itself in World War I's aftermath and how, under the leadership of individuals such as Hans von Seeckt, it created the doctrinal foundations for the lightning war, or Blitzkrieg, that Germany unleashed during the opening campaigns of World War II. Chapter contents address lessons Germany learned from World War I, such as not achieving decisive victory on the western front; how Von Seeckt emphasized the importance of technical education and officers needing to meet very high educational standards; ways of developing and training the new German military (Reichswehr) whose size was restricted by the victorious Allied Powers; incorporating modern weaponry into this new German military structure; developing an airpower doctrine to accommodate the increasing importance of aviation in military operations; and how collaboration with the Soviet Union helped enhance the emergence of German military power, which would come to devastating fruition during the Nazi era.

Dick, C. J. *Russia's 1999 Draft Military Doctrine.* Camberley, Surrey: Conflict Studies Research Center, 1999. Also available online at http://www.da.mod.uk/colleges/csrc/archive/russia/OB72.pdf/.

Analysis emphasizing how Russia's evolving military doctrine was primarily defensive in nature and reflective of its apparent establishment of a democratic state. Some attributes of this Russian doctrine include recognizing a diminished threat of a world war, including a nuclear war; increasing ethnic nationalism and religious extremism; the proliferation of weapons of mass destruction; and the diminished ability of international organizations such as the United Nations and Organization for Security and Cooperation in Europe to ensure international security.

Specific Russian security concerns cited include intervention in Russian "internal affairs" by outside actors; discrimination against Russian citizens in former Soviet territories; information warfare directed against the Russian federation and its allies; international terrorism; and concern over eastern NATO expansion. Additional characteristics of this nascent Russian military doctrine include the roles of civilian leaders, such as the President, and military leadership, such as the General Staff, in formulating military policy; nuclear weapons and their role in national military strategy; the need for an independent and effective scientific and technological support infrastructure; and the needs for conventional arms exports and powerful allies to counterbalance perceived U.S. military dominance.

Dorman, Andrew M. *Transforming to Effects-Based Operations: Lessons From the United Kingdom Experience.* Carlisle Barracks, PA: U.S. Army War College Strategic Studies Institute, 2008. Also available at http://purl.access.gpo.gov/GPO/LPS90365.

Dorman seeks to describe how the British military has sought to transform itself into a force emphasizing effects-based operations and to assess the results of this shift in emphasis. Subjects addressed within this treatise include areas where the U.S. Army could learn lessons from British policies; areas where the U.S. Army and British Ministry of Defence could develop integrated or comparable standards and doctrines for future alliance/coalition operational transformation; and implications for closer U.S. Army cooperation with the UK.

The initial section reviews evolution in British defense policy since the Cold War, while evaluating how much this evolution has produced an effects-based approach. Subsequent sections examine post–Cold War British operational experience, including analysis of lessons learned and British experience working with allies, British capability development through doctrinal and acquisition strategies, and implications of these findings for the U.S. Army, including recommendations. Examples of these findings and recommendations include improving joint cooperation between British air and naval forces and UK after-action reports that place excessive emphasis on what went wrong and not enough on what went right during individual military operations.

Dunnavent, R. Blake. *Brown Water Waterfare: The U.S. Navy in Riverine Warfare and the Emergence of a Tactical Doctrine, 1775–1970.* New Perspectives on Maritime History and Nautical Archaeology. Gainesville: University Press of Florida, 2003. ISBN 0-8130-2614-8.

Dunnavent emphasizes that few nations have conducted as extensive riverine military operations as the United States, and stresses how important this brand of military warfare has been to the United States and to national military strategy. He describes the importance of riverine operations in the Revolutionary War, the War of 1812, the Second Seminole War, the Mexican-American War, conflict on the Rio Grande during the 1870s, the Philippine war of the early 20th century, operations in China during the 1920s and 1930s, and during the Vietnam War. Operations during this last conflict led the Marine Corps to develop in April 1966 the Fleet Marine Force Manual (FMFM) 8–4, *Interim Doctrine Riverine Operations,* and two years later the Navy would adopt Naval Warfare Publication (NWP) 21(A), *Doctrine for Riverine Operations,* to provide brown water sailors with guidance for conducting operations in riverine environments.

Echevarria, Antulio J., II. *After Clausewitz: German Military Thinkers before the Great War.* Modern War Studies. Lawrence: University Press of Kansas, 2000. ISBN: 0-7006-1071-5.

Echevarria seeks to analyze the theoretical works published by German military authors prior to World War I. The initial chapter seeks to examine how the stress soldiers experienced during combat increased due to firepower advances; how this would bring these troops to a psychological breaking point more quickly; and that battlefield commanders would have to change their infantry, cavalry, and artillery attack strategies because of this accelerated rate of soldiers' psychological collapse.

Later chapters discuss solutions developed to attempt to resolve this crisis, such as debate between those favoring *Normaltaktik* (standardized tactics) and *Auftragstaktik* (mission or task-oriented tactics); how the increasing effectiveness and lethality of firepower technology raised disconcerting questions about the resilience of modern and urban recruits and how German military writers struggled to resolve this dilemma.

Subsequent chapters describe how these military writers reacted to battlefield developments of the Boer and Russo-Japanese wars; how these conflicts seemed to indicate the increasing importance of breakthrough military operations and attacks against fortified positions in emerging military conflicts; the importance of integrating new technologies such as machine guns and aircraft into future military operations; and how these and other technologies influenced fundamental battle conceptions before World War I. Analysis is also presented regarding how American, British, French, and Russian military writers also grappled with these theoretical and doctrinal matters during the years leading up to World War I.

Farrell, Theo and Terry Terriff, eds. *The Sources of Military Change: Culture, Politics, Technology.* Making Sense of Global Security Series. Boulder, CO: Lynne Rienner Publishers, 2002. ISBN: 1-55587-975-6.

This collection of essays examines how militaries have sought to incorporate change into their doctrinal and operational practice. Examples of topics covered in these geographically and historically dispersed essays include how western military models were spread and incorporated into societies as diverse as Ottoman Turkey and Meiji Japan; how the Irish military incorporated British and other global military influences into its operational activities from 1922 to 1942; how U.S. thinking influenced NATO military change from 1989 to 1994; and changes in U.S. military strategic thinking from 1963 to 1988.

Political themes covered here include U.S. military responses to post–Cold War missions and Russian military reform during this time period. Technological military changes that receive scrutiny include the historical evolution of tanks in British military thinking; how technological advances and evolution have influenced recent U.S. military thinking; and the increasing role of information technology as a sculpting force of military doctrinal thinking.

Gray, Colin S. *Weapons Don't Make War: Politics, Strategy, and Military Technology.* Modern War Studies. Lawrence: University Press of Kansas, 1993). ISBN: 0-7006-0559-2.

This work focuses on the relationship between political and military policy, military strategy, and weapons. Topics addressed in specific chapters include relationships between or among offensive or defensive strategies; relationships between policy, strategy, and purportedly offensive or defensive weapons; examining policy, weapons, and alleged arms races; reviewing policy strategy and the weapons acquisition process; investigating policy, strategy, and defense planning for uncertainty; scrutinizing policy, strategy, and arms control; and investigating the historical record of connections between policy, strategy, and weapons during the nuclear era. Gray urges that readers be particularly cautious about trying to extract extreme conclusions about how important the nuclear revolution is in assessing military strategy.

Habeck, Mary. *Storm of Steel: The Development of Armor Doctrine in Germany and the Soviet Union, 1919–1939.* Cornell Studies in Security Affairs. Ithaca NY: Cornell University Press, 2003. ISBN: 0-8014-4074-2.

Habeck seeks to review the development of German and Soviet armor doctrine during this interwar period, with particular emphasis placed on how General Heinz Guderian (1888–1954), Marshall Mikhail Tukhachevsky (1893–1937), and others developed their country's military armor doctrines.

Chapter contents address how both countries sought to incorporate embryonic armor technology and doctrine into their military forces following World War I; debates within both militaries over the mechanization of warfare; early German-Soviet armor doctrine collaboration in the late 1920s; divergence in the armor doctrinal practices of these militaries by the mid 1930s; how both militaries' armor doctrine was tested by German operations during the Spanish Civil War and Soviet operations against Japanese forces in east Asia; and how their armor doctrines were further refined by German operations against Poland and Soviet operations against Finland during the opening months of World War II. These preliminary operations and doctrinal knowledge base would be put to their ultimate test and battlefield application during the titanic German-Soviet confrontation in World War II.

Honna, Jun. *Military Doctrines and Democratic Transition: A Comparative Perspective on Indonesia's Dual Function and Latin American National Security Doctrines.* Canberra: Australian National University, Dept. of Political and Social Change, Research School of Pacific and Asian Studies, 1999. ISBN: 0-731-52676-7.

This dense, theoretical treatise examines how Indonesian and Latin American military doctrines coped with their countries' national political transitions from military rule to civilian democratic structures. Within the Indonesian and Latin American cases, with the latter incorporating countries such as Argentina, Brazil, Chile, and Peru, there was a preexisting doctrinal belief that militaries in these countries equate their fortunes with those of the state.

These militaries' somewhat successful transition to adopting decreasingly influential roles in national political life required numerous painful steps. These included reducing their professional spheres of competence to military operational matters, revising how they envisioned nationalism, accepting the idea that political conflicts are normal and necessary for political stability, and institutionalizing civil-military collaboration in formulating national security policy.

Hough, M., and L. Du Plessis, eds. *Selected Military Issues with Specific Reference to the Republic of South Africa.* Pretoria: University of Pretoria Institute For Strategic Studies, 2001. Ad Hoc Publication No. 38. ISBN: 1-8685-4416-8.

This collection of essays presents a historical review of South African military doctrinal thinking at the beginning of the 21st century. Subjects addressed within this work's six chapters include South African armed forces doctrinal development until the 1980s; national military doctrine since 1994; South African warfighting principles in 2001 in comparison with American and British principles at that same time; the South African government's process for planning military interventions; South African Army combat readiness; and the importance of morale and discipline within South Africa's army as it seeks to meet national security objectives.

A particularly useful case study is provided of South Africa's September 1998 military intervention in Lesotho, which did not achieve optimal success due to poor intelligence, planning, and deployment decisions.

Kilcullen, David. *The Accidental Guerrilla: Fighting Small Wars in the Midst of a Big One.* New York: Oxford University Press, 2009. ISBN: 978-0-19-536834-5.

Kilcullen, a prominent Australian counterinsurgency expert who has advised the U.S. State Department and General David Petraeus and helped implement the 2007 Iraq surge strategy, presents his views on how to fight effective counterinsurgency campaigns. Kilcullen describes "accidental guerillas" as individuals in various areas such as Pakistan who end up fighting Western military forces because of the presence of these forces in their homelands as part of larger military campaigns and emphasizes that these forces can be galvanized by high-tech and internationally oriented ideologues such as al Qaida.

This work examines how such insurgencies have played out in locations as diverse as Afghanistan, Indonesia, Iraq, and Pakistan. His recommendations for Western success in such counterinsurgency campaigns include keeping existing terrorists off balance through strategic disruption, flexible Western tactical responses to continually shifting battlefield and political conditions, providing multifaceted assistance to societies struggling with insurgencies by enhancing local institutions, building and maintaining trust among indigenous populations, and establishing virtue, moral authority, and credibility with these populaces to lessen the appeal of insurgent forces.

Kugler, Richard L. *NATO's Future Conventional Defense Strategy in Central Europe: Theater Employment Doctrine for the Post–Cold War Era.* Santa Monica, CA: Rand Corporation,

1992. ISBN: 0-8330-1188-X. Also available at http://rand.org/pubs/reports/2007/R4084.pdf.

This study, prepared for the U.S. Army, sought to examine NATO's Central European conventional defense outlook in light of the Soviet Union's collapse. It places particular importance on analyzing how NATO employs its battlefield military forces to obtain goals and how the alliance can achieve its goals in light of the then-emerging era of lower military preparedness. It also maintains that German reunification places NATO force structure further east and produces major upheaval in alliance defense planning practices.

The report contents address historical, current, and possible future Central European defense environments, a historical review of linear defense, and a discussion of how such defense strategy can be successfully limited with lower force levels in light of the Soviet Union's demise.

Li, Xiaobiao. *A History of the Modern Chinese Army.* Lexington: University Press of Kentucky, 2007. ISBN: 978-0-8131-2438-7.

This work seeks to examine how the Chinese Army (known as the People's Liberation Army (PLA)) has evolved from a peasant-based labor-intensive military to an increasingly professionalized force desirous of winning technologically intense military conflicts. It focuses on changes and transformations in the PLA's evolution from 1949 to 2002, emphasizing historical trends and development in Chinese military practice until the 1949 Communist Revolution; Chinese military force modernization as a result of the Korean War; how Soviet aid and the 1954–1955 Taiwan Strait crisis influenced emerging PLA military doctrine; the development of a strategic nuclear weapons program between 1955–1964; China's involvement in Vietnam; border conflict with the Soviet Union; the tumultuous upheaval caused by the Cultural Revolution; military modernization begun under Deng Xaioping up to the 1989 Tiananmen Square massacre; and military reforms and events of the 1990s, including the 1996 launching of missiles across the Taiwan Strait.

Li also examines the PLA's commercial activities, its interest in space as a venue for military operations, demographic developments in China that affect the composition of its military forces, and how problems such as unemployment, limited natural resources, rising energy costs, and a weak national financial system may affect the PLA in the future.

Lockwood, Jonathan Samuel, and Kathleen O'Brien Lockwood. *The Russian View of U.S. Strategy: Its, Past, Its Future.* New Brunswick: Transaction Publishers, 1993. ISBN: 1-560-00031-7.

Analysis examining how the Soviet Union viewed U.S. military strategy from approximately Stalin's death in 1953 until its collapse in 1991. The Lockwoods place particular emphasis on the importance of "mirror imaging," in which the Soviets believed that their attitudes on military issues adhered to those of U.S. military policymakers. They also focus on how the Soviets used disinformation about U.S. military strategies to influence world opinion to more positively view Soviet military strategy and policy.

Initial sections of this work address the development and evolution of U.S. and Soviet military doctrine, including Soviet views of U.S. massive retaliation nuclear doctrine. Later chapters address how the Soviets responded to U.S. flexible response nuclear doctrine during the 1960s; Soviet reactions to Nixon Administration military doctrinal pronouncements, differences between Soviet propaganda on U.S. military policies and their actual views; how the Soviets viewed the emergence of the Reagan Administration's Strategic Defense Initiative; and how U.S. military strategy should respond to the Soviet Union's collapse, with particular emphasis on the importance of ballistic missile defense.

Mader, Markus. *In Pursuit of Conceptual Excellence: The Evolution of British Military-Strategic Doctrine in the Post-Cold War Era, 1989–2002*. Bern, Switzerland: Lang, 2004. ISBN: 0-8204-7032-5.

Mader seeks to analyze British military doctrinal development efforts between 1989 and 2002 in this update of his doctoral dissertation. He places particular emphasis on the increasing institutional relevance of doctrine within the British military and how Britain's post–Cold War military strategy is being expressed in doctrine. This work is broken up into two parts, with Part I emphasizing the reemergence of conventional military power and single-service doctrinal developments from 1989 to 1996, and Part II examining how post–Cold War military strategy developments from 1996 to 2002 are placing increasing emphasis on joint doctrine.

Specific topics of individual chapters within these units include how land power is leading doctrinal development toward a capability-based army; the importance of maritime power projection in emerging British military doctrine; how British peacekeeping operations are affecting military doctrinal thought; and how asymmetric conflicts after 9/11 are compelling the British military to incorporate doctrine for this kind of conflict into national military strategy.

Mader contends that Britain's post–Cold War doctrinal development has been strongly influenced by the United States; that developments such as the RMA have profoundly sculpted this emerging British military doctrine; and that ongoing military operations in Afghanistan and Iraq will influence evolving British military doctrinal thinking for the foreseeable future.

Mandeles, Mark David. *Military Transformation Past and Present: Historic Lessons for the 21st Century*. Westport, CT: Praeger Security International, 2007. ISBN: 978-0-275-99190-3.

Mandeles provides assessments of historical and contemporary examples of military transformation. Themes examined in individual chapters include transformation in the post–Civil War by the U.S. Army and Navy; innovations in the interwar period by Army and Navy aviation forces; problems experienced by the Marine Corps and the British Royal Marines in developing amphibious operations;

and recent Navy efforts to develop and exploit concepts and technologies for cooperative engagement capability in areas such as aircraft and ship interoperability.

Mandeles concludes that the criticism inherent in democracies enables military organizations to function more effectively; that improved military capabilities occur due to critical observation and conscious effort to improve the ability of Department of Defense (DOD) entities to identify and correct errors; that participation of senior leadership in a multi-organizational environment is an intentionally strategic choice that increases the probability of errors being identified and removed from acquisition programs, doctrine, and operational concepts; and that robust organizational methods for identifying and eliminating error are conducive to the United States gaining significant combat advantage in future military operations.

Menning, Bruce. *Bayonets before Bullets: The Imperial Russian Army, 1861–1914.* Indiana-Michigan Series in Russian and East European Studies. Bloomington: Indiana University Press, 1992. ISBN: 0-253-33745-3.

This history seeks to analyze military reforms made by Russia's Tsarist military between the Crimean War of the 1850s and World War I. Particular emphasis is placed on reforms implemented in response to lessons learned during the 1877–1878 Turkish War and the 1904–1905 war against Japan. Readers are introduced to military leaders like Dimitri Miliutin (1816–1912), Mikhail Dragomirov (1830–1905), and Alexei Kuropatkin (1848–1925), who sought to make Russia's military forces more capable of meeting their country's national security needs. Conflicts between Tsarist officials and military officers over whether to implement military doctrinal, training, and tactical reforms are analyzed. Menning concludes that Tsarist military practices such as rationalizing success and failure were also adopted by the Communists, who ultimately incorporated Tsarist strategy, operational art, and tactics into their own military doctrine.

Merom, Gil. *How Democracies Lose Small Wars: State, Society, and the Failures of France in Algeria, Israel in Lebanon, and the United States in Vietnam.* New York: Cambridge University Press, 2003. ISBN: 0-521-80403-5 (cloth) and 0-521-00877-8 (pbk).

Merom examines unsuccessful U.S., French, and Israeli experiences in fighting counterinsurgency wars in Vietnam, Algeria, and Lebanon and the factors he considers to be crucial reasons for these defeats. He contends that modern democracies fail in such wars because they are incapable of finding a balance between expedient and moral tolerance of wartime costs. Merom maintains this occurs when a critical minority in these societies shifts the center of gravity from the battlefield to the marketplace of ideas. This minority, which he says is derived from the educated middle-class, despises the brutality necessary for effective counterinsurgency, while also refusing to accept the casualties necessary to successfully conduct counterinsurgency operations. Consequently, governmental institutions further contribute to failure by resorting to harsher behavioral patterns in battlefield operations to overcome their domestic political problems.

Additional observations include recognizing that democracies can effectively adapt to battlefield conditions in these conflicts; that opposing dictatorial leaders may mistakenly calculate the willingness of democratic countries to engage in such wars; and that decisions by democratic militaries to scale down or withdraw from conflicts does not necessarily mean those conflicts will end.

Nagl, John A. *Learning to Eat Soup with a Knife: Counterinsurgency Lessons from Malaya and Vietnam.* Chicago: University of Chicago Press, 2005. ISBN: 978-0-226-56770-9.

This work, viewed by some military doctrine specialists as having near canonical authority, addresses the belief that armies are only prepared to fight previous wars by examining how armies can adapt to changing circumstances during conflicts for which they were initially unprepared. Written by a current military officer, this work examines how counterinsurgency doctrine was developed and practiced during the Malayan Emergency (1948–1960) and during the Vietnam War (1950–1975).

Nagl believes that organizational culture is a key variable enabling militaries to adapt and learn from unexpected conditions, which he believes the British were more successful at doing in Malaysia than the United States was in Malaysia. He also believes the U.S. military should modify its doctrine for low intensity conflict to make doctrinal development a continually evolving group of theoretical guidelines; establish a systemic assessment process to facilitate current doctrinal assumption validity; develop efficient processes for acquiring organizational consensus on emerging doctrines; establish effective practices for rapidly disseminating this doctrine to field units; welcome civilian leadership's inquiries about military capacity and doctrinal appropriateness for military institutions; and view doctrine as a way of inquiring about military effectiveness for potential threats and challenges.

Ng, Ka Po. *Interpreting China's Military Power: Doctrine Makes Readiness.* London: Frank Cass, 2005. ISBN 0-7146-5548-1.

This work examines factors influencing the development of Chinese military doctrine. It begins by acknowledging how the lack of Chinese transparency about their military policy makes conducting research on China's military more complicated. Ng emphasizes how Chinese military strategy has oscillated between conducting local war and total war, with the latter representing existential threats to Chinese national survival. Recent years have seen local war assume preeminence in Chinese military doctrine as China has developed a more professional and technologically oriented military to meet national security objectives. The author believes that a doctrine-based concept of military readiness is most suitable for interpreting Chinese military policy.

Posen, Barry. *The Sources of Military Doctrine: France, Britain, and Germany between the World Wars.* Cornell Studies in Security Affairs. Ithaca, NY: Cornell University Press, 1984. ISBN: 0-8014-1633-7.

This treatise examines the bureaucratic, political power, technological, and geographic influences sculpting national military doctrine, with particular emphasis on French, British, and German military doctrine during the interwar years. Additional emphasis is placed on successful applications of military doctrine during this period, such as the German Blitzkrieg and British air defense system, as contrasted with the failure of the French Army's defensive doctrine as epitomized by the Maginot Line.

Additional topics analyzed by Posen include the importance of offensive, defensive, and deterrent military doctrine characteristics; the roles played by organization theory and balance of power theory in determining interwar French, British, and German military doctrine; that military organizations dislike deterrence doctrines because determining how to break national will is an inherently political task; and that military organizations prefer offensive doctrines since they are likely to increase organizational size and wealth while also reducing external uncertainty if unexpected events such as huge losses or partial defeats in military operations occur.

Posen concludes by stressing how powerful political pressures and technological realities can favor offensive forces and doctrine and emphasizing the importance of opposing politico-military forces placing some restraints on their military competition.

Rose, John P. *The Evolution of U.S. Army Nuclear Doctrine, 1945–1980*. Boulder, CO: Westview Press, 1980. ISBN: 0-86531-029-7.

This work seeks to explain the origins and evolution of U.S. army nuclear doctrine, predicated on the conviction that the United States must be prepared to develop the techniques necessary to fight successfully in a nuclear combat environment. Early sections of this work cover the historic development and evolution of U.S. and Soviet nuclear doctrine; images and realities of nuclear weapons, including the demonstration that militaries have found ways to defend against new military technologies; and nuclear military theory during the 1950s and 1960s and how decreasing emphasis on nuclear weapons in the battlefield during the latter part of this time period resulted in stagnating nuclear strategic thinking.

Later chapters address nuclear doctrinal developments in the Army's educational system; Soviet doctrinal concepts and strategy; constraints on U.S. nuclear battlefield doctrine, such as restrictions on the ability to use nuclear weapons; and the need for the military to incorporate offensive operations into its nuclear warfighting doctrine.

Vlakancic, Peter J. *Marshall Tukhachevsky and the "Deep Battle": An Analysis of Operational Level Soviet Tank and Mechanized Doctrine, 1935–1945*. Land Warfare Papers #14 Arlington, VA: Institute of Land Warfare, Association of the United States Army, 1992.

Vlakancic seeks to examine how Tukhachevsky's doctrine of *gluboky boi,* or "deep battle," influenced Soviet armored military doctrine from 1935 to 1945.

Gluboky boi called for a four-echelon offensive, in-depth strategy focusing on aircraft gaining aerial superiority and bombing enemy positions, unleashing shock groups consisting of tanks, infantry, and artillery to punch a hole in enemy lines, mechanized units assertively exploiting these successes by driving for the enemy's rear and encircling hostile units and infrastructures, and reserves following the third echelon to consolidated its advances.

This doctrine received initial success when it was solidified into Soviet doctrine from 1935 to 1937, experienced setbacks and stagnation following Tukachevsky's 1937 execution and initial defeats in World War II, and ultimately experienced rebirth and success due to Soviet victories achieved using its tenets during the final years of World War II.

Weigley, Russell F. *The American Way of War: A History of United States Military Strategy and Policy*. Bloomington: Indiana University Press, 1977. ISBN 0-253-28029-X.

This classic analysis of American military thinking divides the development and evolution of such thought into five distinct chronological periods. The first, covering 1775–1815, describes how George Washington (1732–1799) and Nathaniel Greene (1742–1786) sought to be effective war fighters with limited material resources and how Federalist and Jeffersonian political factions viewed military strategy.

A second section examines America's emergence as a military power from 1815 to 1890, placing emphasis on the role of figures such as Winfield Scott (1786–1866); the Civil War and Indian wars as serving as fulcrums for developing U.S. military thinking; and the intellectual importance of Dennis Hart Mahan (1802–1871) and Henry Wager Halleck (1815–1872) in developing uniquely American theories of military strategy.

Section three describes the United States' rise to world power from 1890 to 1941 and the role played in this by naval strategists such as Stephen B. Luce (1827–1917) and Alfred Thayer Mahan (1840–1914). The influence of Ulysses Grant (1822–1885) and Mahan are used to describe U.S. European and Asian military strategies during World War II and the influence of the nuclear revolution and Vietnam War in shaping more recent U.S. military strategy and policy is also covered.

Welburn, Mark Christopher John. *The Development of Australian Army Doctrine, 1945–1964*. Canberra Papers on Strategy and Defense No. 108. Canberra: Australian National University, Strategic and Defense Studies Centre, Research School of Pacific and Asian Studies, 1994. ISBN: 0-731-52106-4.

Welburn contends that the Australian Army was initially dependent on other countries, particularly Great Britain, for developing its military doctrine, but that during the two decades after World War II Australian land forces developed a doctrine derived from competing strategic interests and other countries' doctrines, and by emphasizing small-unit operations.

The work's contents describe how events such as the fall of Singapore and the commitment of Australian forces to fight in New Guinea helped lessen Australian reliance on British military doctrine; how the interim five years of peace before the outbreak of the Korean War saw limited training of Australian military forces due to postwar draw downs; how the Korean War saw Australia shift its defense emphasis from the Mideast to Southeast Asia; that this shift in geographic emphasis was reinforced by British and Australian counterinsurgency operations in Malaysia; Australian adoption of a U.S. pentropic or five-unit Army operational structure to facilitate the number of simultaneous conflicts it could fight without increasing Army size; and how it would not be until 1965 that the Australian Army would have promulgated a doctrine enabling it to conduct operations in Southeast Asia, such as in the emerging Vietnam War.

Winton, Harold R. *To Change an Army: General Sir John Burnett-Stuart and British Armored Doctrine, 1927–1938*. Modern War Studies. Lawrence: University Press of Kansas, 1988. ISBN: 0-7006-0356-5.

General Sir John Burnett-Stuart (1875–1958) was an important figure in British military history for his advocacy that British forces incorporate armor into national military doctrine and strategy. Chapters within this work address British military reform from 1870 to 1925; Burnett-Stuart's military education; the emergence of mechanization and the birth of British armor doctrine during the 1920s and 1930s; how Burnett-Stuart's experiences as commander of British military forces in Egypt during the 1930s increased his advocacy of armored warfare; and how British armored doctrine had, by the time of Burnett-Stuart's 1938 retirement, surpassed American and Soviet armored doctrine while lagging behind German armored doctrine.

Zisk, Kimberly Marten. *Engaging the Enemy: Organization Theory and Soviet Military Innovation, 1955–1991*. Princeton: Princeton University Press, 1993. ISBN 0-691-06982-4.

Zisk examines whether military organizations value prestige and organizational stability above other factors and whether they tend to innovate only when they or close allies suffer battlefield defeat, which compels them to adopt new doctrines by civilian intervention into the military doctrine formulation process. Her work examines Soviet military doctrinal innovation in the post–Stalin era.

Zisk concludes that professional military officers are aware of changes occurring in military doctrines and the force postures of potential future enemies; that not all officers from particular service branches act from traditionalist calculations of organizational interest, with some of these individuals being more likely to propose or adopt innovative policy ideas; and that civilian intervention into military doctrinal formulation can take multiple forms and be accompanied by variant levels of bureaucratic contentiousness and organizational hostility.

Engaging the Enemy analyzes Soviet reactions to American or NATO military policy changes such as the 1960s Flexible Response doctrine, the 1974 Schlesinger Doctrine of limited nuclear options, and the combined U.S. adoption of AirLand Battle doctrine and NATO doctrine of Follow-On Forces Attack in the early 1980s.

Note

1. For examinations of the role of the scholarly book in academic literature, see Franklin H. Silverman, *Publishing for Tenure and Beyond* (Westport, CT: Praeger, 1999); John B. Thompson, *Books in the Digital Age: The Transformation of Academic and Higher Education Publishing in Britain and the United States* (Cambridge, UK: Polity, 2005); Kenneth T. Henson, *Writing for Publication: Road to Academic Advancement* (Boston: Pearson/Allyn and Bacon, 2005); Albert N. Greco, Robert M. Wharton, and Hooman Estelami, "The Changing Market for University Press Books in the United States: 1997–2002," *Journal of Scholarly Publishing* 36, no. 4 (2005): 187–220; and Amy Benson Brown, "Where Manuscript Development Meets Faculty Development," *Journal of Scholarly Publishing* 37, no. 2 (2006): 131–135.

CHAPTER 6

Indexes and Scholarly Journals

An essential component of any area of scholarly research is articles published in scholarly journals. This is true for military doctrine as well as for other subjects. Performing effective scholarly research on any subject involves thoroughly searching for individual journal article citations on this subject, and this is best accomplished by searching print indexes or electronic databases rather than perusing bookshelves for articles. Some indexes are freely available on the Internet and their Web site URLs are listed below. Other indexes are produced by commercial companies and may be available in selected academic and public libraries. An example of a freely available periodical index is the *Air University Library Index to Military Periodicals,* produced by Air University Library at Maxwell Air Force Base in Alabama. This long-standing military science literature index covers 1988–present and is freely accessible at http://purl.access.gpo.gov/GPO/LPS3260.

This index features detailed citations and links to subject headings for additional research. Upon retrieving citations from this and other databases, users will need to check their local libraries to see if they have paper or electronic copies of the articles cited in these resources.

America: History and Life, produced by ABC-CLIO, indexes articles, book chapters, books, and dissertations on American and Canadian history from 1450 to the present. It will be available in many academic libraries and general information on it is available at http://www.abc-clio.com/.

EBSCO's Military and Government Collections is another resource produced by a prominent libraries serial vendor. It provides full-text access to articles from nearly three hundred journals and periodicals, along with numerous pamphlet resources with retrospective coverage that dates back to the mid–1980s. General information on this is accessible at http://www.ebsco.com/.

ABC-CLIO also produces the database Historical Abstracts, which indexes articles, book chapters, books, and dissertations on national and international history outside North America from 1450 to the present. It is available in many academic libraries and general information on it is accessible at http://www.abc-clio.com/.

Published by LexisNexis Inc., the LexisNexis Government Periodicals Index provides access to over 170 U.S. Government periodicals from 1988 to the present. It is available in many medium or large academic libraries and general information can be found at http://academic.lexisnexis.com/online-services/government-periodicals-index.overview.aspx.

Public Affairs Information Service (PAIS) is produced by the Cambridge Scientific Abstracts in Bethesda, MD. Its focus is providing access to scholarly public policy literature from journal articles, books, book chapters, and selected U.S. Government documents. Many academic libraries subscribe to its print or online services and general information on it can be found at http://www.csa.com/factsheets/pais-set-c.php.

The Staff College Automated Periodicals Index (SCAMPI) is produced collaboratively by the Joint Forces Staff College Library, National Defense University Library, and Defense Technical Information Center. It provides bibliographic access to popular and scholarly military publications along with selected public policy institution research reports from 1997–present. SCAMPI is freely accessible at http://www.dtic.mil/dtic/scampi/.

Worldwide Political Science Abstracts (WPSA) is published by Cambridge Scientific Abstracts. It indexes articles from approximately 1,690 political science journals from 1975–present, in addition to some retrospective coverage from 1960–1974. General information on this database is accessible at http://www.csa.com/factsheets/polsci-set-c.php

Scholarly Journals

Many history, military science, and political science journals produce scholarly articles on various aspects of military doctrine and doctrinal thought. Scholarly journals publish articles that have gone through the peer review process in which the journal's editorial board, consisting of experts and scholars in that field, review proposed articles to determine their suitability for publication. Scholarly journals are distributed in print and electronic formats and are available in varying degrees at U.S. and foreign academic libraries. Prevailing practices in academic libraries, however, are emphasizing electronic access and holdings as the preferred method for users to use these resources and for libraries to retain them.[1]

A small number of these journals published by government agencies and nonprofit organizations may be freely available on the Internet. Most of these journals, however, are published by commercial for-profit publishers and are not freely available in print or electronic format. College or university libraries that have print and electronic access to these journals have paid for this access by negotiating contractual agreements with these periodicals publishers. Such agreements may restrict electronic access to these journals to users who are part of a university community, such as faculty and students with university identification numbers. These agreements may stipulate that only computers in the university library or the university's IP range may be used to access electronic journal contents.

A helpful directory of scholarly periodicals is *Ulrich's International Periodicals Directory*. This annual multivolume set, published by R. R. Bowker, is a key source in many academic libraries for locating periodical information.

Two subscription-based projects that provide subscribing academic libraries with numerous electronic journals on various subjects are JSTOR and ExLibris MetaLib. JSTOR provides access to recent and historical issues of scholarly journals in several social science disciplines. Information on JSTOR is accessible at http://www.jstor.org/. ExLibris MetaLib is an international information service provider delivering access to electronic journal articles in multiple subjects at many academic and research institutions. General information on this service is accessible at http://www.exlibrisgroup.com/category/MetaLibFAQ.

An increasingly important aspect of scholarly journal publishing is the growth of the open access movement. This initiative seeks to provide a counterpoint to the sometimes restrictive access policies commercial publishers place on their works. Open access movement proponents advocate that scholars publish their research in journals that do not have restrictive public access policies or do not charge high and continually rising institutional subscription prices for their journals.[2] Information on this increasingly important scholarly publishing movement can be found at http://www.publicknowledge.org/issues/openaccess/.

The following section is a representative sampling of important scholarly journals that produce articles on military doctrine. The information provided includes the journal's name, publisher, paper and electronic International Standard Serial Numbers (ISSN), publication frequency and history, and general information about its accessibility, including a URL if it is freely available to the general public.

African Security Review

African Security Review is published by the Institute for Security Studies (ISS) in Pretoria and Cape Town, South Africa, with additional facilities in Nairobi, Kenya and Addis Ababa, Ethiopia. It is published quarterly, its ISSN is 1024-6029, and it has been published since 1992. General information on *African Security Review* and access to its contents is accessible through the ISS Web site (http://www.iss.co.za/). Sample articles on military doctrine include "A Pan-African Army: The Evolution of an Idea and Its Eventual Realisation in the African Standby Force" (2006); "A Critical Analysis of Africa's Experiments with Hybrid Missions and Security Collaboration" (2007); "A Plan for Military Intervention in Darfur" (2007); and "The African Union's Evolving Role in Peace Operations: The African Union Mission in Burundi, the African Union Mission in Sudan, and the African Union Mission in Somalia" (2008).

Air and Space Power Journal

Air and Space Power Journal or *Aerospace Power Journal* is the U.S. Air Force's preeminent professional military journal, and it is produced quarterly by Air University at Maxwell Air Force Base, AL. Its ISSNs are 0897-0823 and 1555-385X, it has been published since 1947. Current and many historical issues are freely accessible to the public at http://purl.access.gpo.gov/GPO/LPS25494. Sample articles focusing on military doctrine as applied to aerospace forces include "Of Trees and Leaves: A New View of Doctrine" (1982); "The Problem with Our Air Power Doctrine" (1992); "Air-Minded Considerations for Joint Counterinsurgency Doctrine" (2007); "Exposing the Information Domain Myth: A New Concept for Air Force and Information Operations Doctrine" (2008); and "Integrating Weather in Net-Centric Warfare: A Case for Refocusing Human Resources in Air Force Weather" (2008). This journal is an essential resource for those studying the historical development and evolution of U.S. aerospace military doctrine.

Armed Forces and Society

Armed Forces and Society is produced by the InterUniversity Seminar on Armed Forces and Society (IUS) at Loyola University–Chicago. It is published quarterly by Sage Publications. It has been published since 1975 and its paper and electronic ISSNs are 0095-327X and 1556-0848. General information on the journal can be found at http://www.sagepub.com/journalsIndex.nav. Examples of pertinent articles include "The Israel Defense Forces (IDF): From a 'People's Army' to a 'Professional Military'—Causes and Implications" (1995); "Israel's National Security Doctrine under Strain: The Crisis of the Reserve Army" (2002); "India's Nuclear Doctrine and Command Structure: Implications for Civil-Military Relations in India" (2007); and "The Competing Claims of Operational Effectiveness and Human Rights in the Canadian Context" (2008).

Australian Army Journal

Australian Army Journal is published by the Australian Army's Land Warfare Studies Centre in Duntroon, Australia. It is published three times a year, its ISSN is 1448-2443, and it has been published since 2003. It is freely available to the public at http://www.defence.gov.au/army/lwsc/Australian_Army_Journal.htm.

Pertinent sample articles include "Rethinking the Basis of Infantry Close Conflict" (2003); "The Australian Defence Force and the Continuing Challenge of Amphibious Warfare" (2004); "Uninhabited Combat Aerial Vehicles and the Law of Armed Combat" (2006); and "Character and the Strategic Soldier: The Development of Moral Leadership for the All Corps Soldier Training Continuum" (2007).

Australian Defence Force Journal

Australian Defence Force Journal is published bimonthly by the Australian Department of Defence. Its ISSN is 1444-7150, it has been published since 1976, and articles from 1997–present are freely available at http://www.defence.gov.au/dfj/. Representative sample articles include "Psyops beyond 2000: Coordinating the Message" #125 (1997); "The Relevance of a Concept of Cooperative Security" #140 (2000); "International and Australian Pre-Emption Theory" #163 (2003); and "The Clash of Cultures: Command and Control in Joint Warfare" #174 (2007). Both *Australian Army Journal* and *Australian Defence Force Journal* provide excellent insights into Australian military thinking on military doctrine issues.

Canadian Army Journal

Canadian Army Journal is published quarterly by the Canadian military's Land Force Command. It began publishing in 1947, and it has been published online since 1998. Its ISSN is 1713-773X, and general information on the journal and access to its contents are available at http://www.army.forces.gc.ca/CAJ/. Representative articles include "From the Directorate of Army Doctrine Firepower: A Primer for the New Manual" (1999); "The Urban Web: An Operational Concept for Offensive Operations in the Urban Sprawl of the 21st Century" (2004); "The Role of the Artillery in Afghanistan" (2007); and "Learning on the Run: Company Level Counter-Insurgency in Afghanistan" (2008).

Canadian Military Journal

Canadian Military Journal is published by Canada's Department of National Defence. It has been published quarterly since 2000, its ISSNs are 0008-4468 and 1494-465X, and its contents are accessible at http://www.journal.forces.gc.ca/. Representative articles include "2020 Vision: Canadian Forces Operational-Level Doctrine" (2001); "The Evolution of the Canadian Approach to Joint and Combined Operations at the Strategic and Operational Level" (2002–2003); "The New Political Reality of Pre-Emptive Defence" (2005); and "Towards a More Strategic Future?: An Examination of the Canadian Government's Recent Defense Policy Statements" (2006).

Defense & Foreign Affairs Strategic Policy

Defense & Foreign Affairs Strategic Policy has been published 10 times a year by the International Strategic Studies Association since 1972. Its ISSN is 0277-4933, and general information on this periodical can be found at http://www.strategicstudies.org/.

Examples of its military doctrine articles include "Lessons of Iraq War: A Pivotal War: Strategically, Tactically, Technologically" (2003); "Iranian, Wahhabist, and Syrian Patterns Clarify" (2005); "Learning from History about Future Options for Space" (2007); and "The Strategic-Tactical Relationship: For Want of a Nail" (2008).

European Security

European Security is published quarterly by Taylor and Francis. It has been published since 1992, and its paper and electronic ISSNs are 0966-2839 and 1746-1545. General information is available at http://www.tandf.co.ul/journals/titles/09662839.asp. Representative articles it has published on military doctrine include "National Interests and Geopolitics: A Primer on 'The Basic Provisions of the Military Doctrine of the Russian Federation'" (1995); "Evidence of Russia's Bush Doctrine in the CIS" (2005); "Was the U.S. Invasion of Iraq NATO's Worst Crisis Ever? How Would We Know? Why Should We Care?" (2007); and "Superficial Not Substantial: The Ambiguity of Public Support for Europe's Security and Defense Policy" (2007).

International Security

International Security is produced at the Belfer Center for Science and International Affairs (BCSIA) at Harvard University's John F. Kennedy School of Government. It has been published quarterly by Massachusetts Institute of Technology (MIT) Press since 1976, and its paper and electronic ISSNs are 0162-2889 and 1531-4804. General information on *International Security* can be found through the Belfer Center Web site (http://belfercenter.ksg.harvard.edu/) and the publisher's Web site (http://mitpressjournals.org/loi/isec). Pertinent *International Security* articles on military doctrine include "The Rise and Fall of Navies in East Asia: Military Organizations, Domestic Politics, and Grant Strategy" (2002); "State Militarism and its Legacies: Why Military Reform Has Failed in Russia" (2004; "Friends Like These: Counterinsurgency and the War on Terrorism" (2006); and "A Cold Start for Hot Wars: The Indian Army's New Limited War Doctrine" (2007/2008). Articles such as these demonstrate why this journal is one of the most important ones in studying national security policy.

Joint Force Quarterly

Joint Force Quarterly is published quarterly by National Defense University. It has been published since 1993, and its ISSNs are 1070-0692 and 1559-6702. Journal contents are freely available at http://www.dtic.mil/doctrine/jel/jfq_pubs/. Articles on military doctrine in this journal include "A Primer on Naval Theater Air Defense" (1996); "Civil-Military Operations: Joint Doctrine and the Malayan Emergency" (2002); "Global and Theater Operations Integration" (2007); and "Attacking Al Qaeda's Operational Centers of Gravity" (2008).

Journal of American History

Journal of American History is one of the premier scholarly journals of U.S. History. It is published by the Organization of American Historians, and has been published quarterly since 1914. General information on the journal is accessible at http://www.indiana.edu/~jah/. Sample military doctrine articles published here include "American Atomic Strategy and the Hydrogen Bomb Decision" (1979); "United States Military Strategy in South Asia: Making a Cold War Commitment to Pakistan, 1947–1954" (1988); "This is the Army: Imagining a Democratic Military in World War II" (1998); and "9/11, the Great Game, and the Vision Thing: The Need for (and Elements of) a More Comprehensive Bush Doctrine" (September 2002 Special Issue).

Journal of Cold War Studies

Journal of Cold War Studies is produced quarterly by Harvard University's Project on Cold War Studies, and it is published by MIT Press. It has been published since 1999, its ISSNs are 1520-3972 and 1531-3298, and additional general information can be found at http://www.fas.harvard.edu/~hpcws/journal.htm and through MIT Press's Web site (http://mitpres.mit.edu/loi/jcws). Sample journal articles on Cold War military doctrinal matters include "The Soviet Military and the Disintegration of the USSR" (2002); "The Nixon Administration, 'The Horror Strategy', and the Search for Limited Nuclear Options, 1969-1972: Prelude to the Schlesinger Doctrine" (2005); "The Cold War Origins of U.S. Central Command" (2006); and "A Most Special Relationship: The Origins of Anglo-American Nuclear Strike Planning" (2007).

Journal of Military History

Journal of Military History is published quarterly by the Society for Military History at Virginia Military Institute in Lexington, VA. It has been published since 1937, its ISSNs are 0899-3718 and 1543-7795 and general information and table of contents from 1997–present are accessible at http://www.smh-hq.org/jmh/. Sample articles from this historical journal on military doctrine include "To Stem the Red Tide: The German Report Series and its Effect on American Defense Doctrine, 1948–1954" (1993); "The Luftwaffe's Army Support Doctrine, 1918–1941" (1995); "The Historiography of Airpower: Theory and Doctrine" (2000); and "Comparing Pearl Harbor and '9/11': Intelligence Failure? American Unpreparedness? Military Responsibility?" (2003).

Journal of Slavic Military Studies

Journal of Slavic Military Studies is published quarterly by Frank Cass. Its ISSNs are 1351-8046 and 1556-3006, it has been published since 1988, and general information on its contents is accessible at http://www.tandf.co.uk/journals/titles/01402390.asp. Pertinent articles include "Russian Nuclear Command and Control: Mission Malaise" (2001); "Soviet Military Doctrine as Strategic Deception: An Offensive Military Strategy for the Defense of the Socialist Fatherland" (2003); "The Serb Guerilla Option and the Yugoslav Wars: Assessing the Threat and Crafting Foreign Policy" (2004); and "The Canadian-Siberian Expeditionary Force, 1918–1919, and the Complications of Coalition Warfare" (2007).

Journal of Strategic Studies

Journal of Strategic Studies is published quarterly by Frank Cass. It is published six times per year, has been published since 1988, and its ISSNS are 0140-2390 and 1743-937X. General information about its contents is available at http://www.tandf.co.uk/journals/titles/01402390.asp. Examples of military doctrine articles in this journal include "Information Capabilities and Military Revolutions: The Nineteenth Century Experience" (2004); "The Israel Defense Forces as an Epistemic Authority: An Intellectual Challenge in the Reality of Israeli-Palestinian Conflict" (2007); "Securing Borders: China's Doctrine and Force Structure for Frontier Defense" (2007); and "Through the Looking Glass: The Soviet Military-Technical Revolution and the American Revolution in Military Affairs (2008).

Korean Journal of Defense Analysis

Korean Journal of Defense Analysis is published by the Korea Institute for Defense Analysis (KIDA) in Seoul. The journal is published quarterly. It has been published since 1989, and its ISSN is 1016-3271. The complete text of journal articles is available from 1999–present through the KIDA Web site (http://www.kida.re.kr/). Pertinent articles include "Nuclear-Armed North Korea and South Korea's Strategic Countermeasure" (2004); "Analyzing South Korea's Defense Reform 2020" (2006); "China's ASAT Test and the Strategic Implications of Beijing's Military Space Policy (2007); and "Playing with Fire: The United States Nuclear Policy toward North Korea" (2007)

Military Intelligence Professional Bulletin

Military Intelligence Professional Bulletin is published by the U.S. Army Intelligence Center at Fort Huachuca, AZ. It is published quarterly, has been published since 1974, and its ISSN is 0026-4024. Issues of this journal from October 2000–present are available at http://purl.access.gpo.gov/GPO/LPS1654. Applicable articles include "Russia's Military Doctrine" (1994); "Transforming the Army for the Next Century—The Future is Here Today! (2000); "Doctrine Corner: U.S. Army Intelligence Center and School Requirements for Lessons Learned" (2003);

"Doctrine Corner: Open-Source Intelligence Doctrine" (2005); and "Priority Intelligence Requirements in Stability and Reconstruction Operations: Doctrine versus Practice" (2007).

Military Review

Military Review is published bimonthly by the U.S. Army Combined Arms Center at Fort Leavenworth, KS. It serves as the Army's principal professional journal and has been published since 1922. Its ISSN is 0062-4148, and access to its contents is provided at http://purl.access.gpo.gov/GPO/LPS53409. Sample articles from this journal, which is a rich resource of U.S. Army military doctrinal thinking, include "Firepower, Attrition, Maneuver—U.S. Army Operations Doctrine: A Challenge for the 1980s and Beyond" (1997); "Integrating Carrier-Based Electronic Attack into Conventional Army Doctrine" (2003); "Engaging Civil Centers of Gravity and Vulnerabilities" (2004); "Army Planning Doctrine: Identifying the Heart of the Problem" (2007); and "FM 3–0 Operations: The Army's Blueprint" (2008).

Military Thought

Military Thought is a Russian journal of military theory and strategy produced by the Russian Federation's Ministry of Defense. It is published quarterly by East View Information Services, and has been published since 1918. Its ISSN is 0236-2058, and general information on it is accessible at http://www.eastview.com/evpj/evjournals_new.asp?editionid=555. Sample articles include "Characteristic Traits of Warfare in Wars and Armed Conflicts in the Last Decade" (2004); "Certain Principles and Problems in Antiamphibious Coast Defense" (2006); "On the Protection of the Tactical Troop Formations in Combined-Arms Combat" (2006); "Russia's Aerospace Journey: The Long Journey in a Maze of Problems" (2007); and "Strategic Nuclear Weapons in Russia's Military Doctrine" (2007).

National Institute of Defense Studies (NIDS) Security Reports (Japan)

National Institute of Defense Studies Security Reports is published by the research branch of the Japanese Ministry of Defense. It English language edition has been published annually since 2000, its ISSN is 1344-1116, and access to its contents can be found at http://www.nids.go.jp/english/. Pertinent articles include "Ocean Peace Keeping and New Roles for the Maritime Force" (2000); "The Nuclear Policy of India and Pakistan" (2003); "The Iraq War, the United Nations Security Council, and the Legitimacy of the Use of Force" (2005); and "Dealing with the Ballistic Missile Threat: Whether Japan Should Have a Strike Capability under its Exclusively Defense-Oriented Policy" (2006).

Naval War College Review

Naval War College Review is published quarterly by the United States Naval War College Press. It has been published quarterly since 1948, its ISSN is 0028-1484, and access to articles from 2004–present and to an index of articles from 1948–present is accessible at http://purl.access.gpo.gov/GPO/LPS17060. Examples of articles on military doctrine from this key journal of navy strategic and operational thinking include "Maritime Geostrategy and the Development of the Chinese Navy in the Early Twenty-First Century" (2006); "A Bi-Modal Force for the National Maritime Strategy" (2007); "Air Force-Navy Integration in Strike Warfare: A Role Model for Seamless Joint-Service Operations" (2008); and "The New Maritime Strategy: A Lost Opportunity" (Spring 2008).

Parameters: U.S. Army War College Quarterly

Parameters: U.S. Army War College Quarterly is published by the U.S. Army War College and serves as an army professional journal. It has been published since 1971, its ISSN is 0031-1723, and access to articles from 1996–present and some articles prior to that is available at http://www.carlisle.army.mil/usawc/parameters/. Examples of the rich corpus of military doctrinal analysis in this journal include "Doctrine is Not Enough: The Effect of Doctrine on the Behavior of Armies" (2000); "Modern War, Modern Law, and Army Doctrine: Are We in Step for the 21st Century?" (2002); "Campaign Design for Winning the War . . . and the Peace" (2005); and "U.S. COIN Doctrine and Practice: An Ally's Perspective" (2007).

Pointer

Pointer is the professional journal of Singapore's armed forces. It is published by that country's Ministry of Defense through the Singapore Armed Forces Technology Institute (SAFTI) Military Institute. It is published quarterly, has been published since 1975, its ISSN is 0217-3956, and issues from 1998–present are freely accessible at http://www.mindef.gov.sg/safti/pointer/.

Examples of military doctrine-related literature in *Pointer* include "Developments Affecting Military Force Planning" (2004); "Connectedness and Cooperation in the 21st Century: The RSAF's Perspective and Practice of Multilateralism" (2005); "Maritime Security: Possibilities for Terrorism and Challenges for Improvement" (2006); and "Networking for Integrated Ground Operations" (2007).

RUSI Journal

RUSI Journal is produced by the British Royal United Services Institute and published by Routledge. It is published bimonthly, has been published since 1858, its ISSNs are 0307-1847 and 1744-3078, and general information is available at http://www.tandf.co.uk/journals/titles/03071847.asp. Individual articles covering military doctrinal topics include "Revisiting Established Doctrine in an Age of Risk" (2005); "Is UK Doctrine Relevant to Global Insurgency?" (2007); "Post-Colonial African Challenges for Peace and Security: The Future of African Military Forces" (2007); and "Learning, Adapting, and Applying U.S. Counter-Insurgency Doctrine and Practice" (2007).

Security Studies

Security Studies is published quarterly by Routledge. It has been published since 1991, its ISSNs are 0963-6412 and 1556-1852, and general information is available at http://www.tandf.co.uk/journals/titles/09636412.asp. Pertinent articles include "Shaping Military Doctrine in France: Decisionmakers between International Power and Domestic Interests" (2001); "Managing Military Transformations: Agency, Culture, and the U.S. Carrier Revolution" (2005); "Norms and Military Power: NATO's War Against Yugoslavia" (2006); "The Preventive War That Never Happened: Britain, France, and the Rise of Germany in the 1930s" (2007); and "Surprise Attacks—Are They Inevitable?: Moving Beyond the Orthodox-Revisionist Dichotomy" (2008).

Small Wars Journal

Small Wars Journal is an online journal produced by former Marine Corps members and run by Small Wars Journal LLC. It seeks to analyze military conflicts and operations in areas such as counterinsurgency, support and stability operations, peacekeeping, noncombatant evacuation, disaster relief, and other related topics. It has been published since 2005. General information on the journal and access to its contents, including articles, blogs, and U.S. military doctrine documents, are available at http://smallwarsjournal.com/. Relevant articles on military doctrine include "Mao in Mufti?: Insurgency Theory and the Islamic World" (2006); "The Marine Corps Small Wars Manual and Colonel C. E. Callwell's Small Wars—Relevant to the Twenty-First Century or Irrelevant Anachronisms?" (2006); "Progressive Reconstruction: Melding Expeditionary Maneuver Warfare with Nation-Building Stability Operations" (2007); "The Political Officer as Counter-Insurgent: Conducting Tactical Politics against Insurgencies" (2007); and "Understanding Iran's Motives in Iraq: The Cost Calculus of External Support" (2007).

Survival: Global Politics and Strategy

Survival: Global Politics and Strategy is produced by the International Institute for Strategic Studies and published by Routledge. It is published quarterly since 1989, its ISSNs are 094-6553 and 1468-2699, and general information on it is available at http://www.tandf.co.uk/journals/titles/00396338.asp. Applicable recent articles include "After the Tests: India's Options" (1998–1999); "The Paradox of Israeli Power" (2004–2005); "Making Strategy: Civil-Military Relations after Iraq" (2006); and "China's Military Space Strategy" (2007).

Notes

1. See N. M. Stanley, "The Case for Acquiring and Accessing Electronic Journals in Libraries," *Collection Management* 19, no. 3/4 (1995): 29–34; Stephen Crothers, Margaret Prabhu, and Shirley Sullivan, "Electronic Journal Delivery in Academic Libraries," *Acquisitions Librarian* 19, no. 37/38 (2006): 15–45; Chandra Prabha, "Shifting From Print to Electronic Journals in ARL University Libraries," *Serials Review* 33, no. 1 (2007): 4–13; Lisa Hanson O'Hara, "Providing Access to Electronic Journals in Academic Libraries: A General Survey," *Serials Librarian* 51, no. 3/4 (2007): 119–128; and Golnessa Galyani Moghaddam, "Archiving Challenges of Scholarly Electronic Journals: How Do Publishers Manage Them?," *Serials Review* 33, no. 2 (2007): 81–90.

2. See Charles A. Schwartz, "Reassessing Prospects for the Open Access Movement," *College and Research Libraries* 66, no. 6 (2005): 488–495; and Emma McCulloch, "Taking Stock of Open Access: Progress and Issues," *Library Review* 55, no. 6 (2006): 337–343.

CHAPTER 7

Grey Literature: Dissertations, Theses, Technical Reports, Think Tanks, and Conference Proceedings

A significant literary corpus for conducting military doctrine research is grey literature. There are many ways to define grey literature and the roles it plays in scholarly literature and research libraries collection development policies.[1] Grey literature normally refers to literature not found in conventional formats such as books, journal articles, government or military documents, or through the print indexes or electronic databases normally used to find conventional scholarly research literature. This chapter will examine literature on military doctrine as appearing in doctoral dissertations, masters' theses, technical reports, and conference proceedings.

Most of this literature will not be freely available on the Internet. Effective access to these information resources will best be provided in academic research libraries that have purchased often expensive commercial databases that provide access to these resources. Descriptions of these databases will be provided later in this chapter. Besides including overviews of these grey literature resource types, this chapter will also include bibliographic citations and annotations for representative samplings of grey literature in these particular genres.

Dissertations and Theses

Doctoral dissertations and masters' theses represent written documentation of their authors' intellectual mastery of various subjects, as well as the successful defense of their findings in oral examinations conducted by their thesis and dissertation supervisors in the process of obtaining their degrees. Writing a thesis or dissertation is an intellectually and physically demanding process that helps enhance the knowledge of intellectual disciplines and branches within these disciplines. A significant body of literature exists on the role of doctoral dissertations in the academic research process.[2]

Once dissertations have been successfully defended, they are eventually deposited in university libraries. In most cases, dissertations are only likely to be housed in libraries at the institutions where they were written. However, some academic research libraries will make efforts to purchase dissertations to enhance the quality of their collection in selected areas. Providing efficient bibliographic access to dissertations and theses has been problematic for academic libraries as various studies document. The Internet's growth has helped improve access to these resources as many libraries have developed digital institutional repositories to provide varying levels of access to theses and dissertations with some success.[3]

Military dissertations and theses have received limited and dated coverage as unique intellectual resources facilitating the sculpting of military knowledge.[4] Military graduate schools, including Air University, the U.S. Army War College, U.S. Army Command and General Staff College, National Defense University, Naval Postgraduate School, and Naval War College, and their organizational components require their attendees to produce scholarly theses or dissertations or comparable high-level analytical written products as part of their degree requirements.[5]

Military officers doing master's and doctoral work at civilian universities produce theses and dissertations on military doctrine and other topics as do their civilian counterparts. There are many ways to access theses and dissertations produced on military doctrine and related subjects. University Microfilms International (UMI), located in Ann Arbor, MI, is a major repository for theses and dissertations. Most theses and dissertations are only available in non-electronic formats, but many are available electronically through UMI's ProQuest Dissertations & Theses (PQDT) service. General information about this service is available at http://www.proquest.com/promos/product/feature01_umi.shtml. This is a paid subscription service and access to it will generally be restricted to academic library users.

There are some additional caveats to consider when trying to locate theses and dissertations. A limited number of universities participate in UMI's theses and dissertation programs, so you cannot be sure that your literature search will retrieve all relevant documents. Only dissertations produced within the last decade or so are likely to be available online through these services. Dissertation authors may choose not to make their dissertations available electronically to PQDT or to make them available for purchase or thru Interlibrary Loan. It is difficult to obtain theses and dissertations from countries outside the United States, Canada, Australia, and the United Kingdom because documents from these countries are not readily available through international bibliographic service providers like UMI or the Online Computer Library Consortium (OCLC).

Useful online repositories to search for and in some cases find the full-text of theses and dissertations include the Theses Canada Portal, produced and maintained by Library and Archives Canada (http://www.collectionscanada.gc.ca/thesescanada/index-e.html), which provides bibliographic citations for Canadian university theses and dissertations from 1965–present; the Australasian Digital Theses Program (http://adt.caul.edu.au/), which is a collaboration of Australian and New Zealand universities that originated in 1998–1999 and that is supported by the Council of Australian University Librarians; the Networked Digital Library of Theses and Dissertations (http://www.ndltd.org/), which consists of various U.S. and European universities whose origins date from 1987–1996 at Virginia Tech University; and Index to Theses (http://www.theses.com/), which indexes and provides limited abstracting and no full text access to theses produced in Great Britain and Ireland since 1716.

The next section of this chapter features a partial selection of theses and dissertations on various aspects of military doctrine from numerous universities over recent decades. Entries will include requisite bibliographic citations and excerpts from the abstracts or summaries of these documents. Readers should check to see if these documents are available electronically through their libraries' database subscriptions or Interlibrary Loan.

These documents were written for various degree programs, represent divergent theoretical and methodological research perspectives, use multifaceted research sources in their bibliographies, and their authors may have gone on to careers in academe or the military. Selection of these resources does not mean the author endorses or opposes the conclusions reached in these documents. Their selection and inclusion in this work illustrates the author's contention that theses and dissertations can be valuable sources for conducting substantive research on the military doctrines of the United States and other countries.

Adams, Thomas Knight. "Military Doctrine and the Organizational Culture of the United States Army." PhD diss., Syracuse University, 1990.

Adams describes U.S. Army doctrine as representing a set of authoritative principles and approved solutions to basic war fighting questions. This treatise emphasizes how Army doctrine remained focused on mass warfare in Europe from the end of World War II until 1989 even though the Army's actual experience involved other warfare forms and geographic locales. Adams argues that the Army failed to adapt its doctrine to technological change occurring during the aforementioned time period and that Army organizational culture and division of professionalization into "political" and "military" spheres of influence make it difficult for the Army to accept political compromise and ambiguity and successfully adapt to emerging forms of military conflict that are frequently morally and politically ambiguous.

Avant, Deborah Denise. "The Institutional Sources of Military Doctrine: The United States in Vietnam and Britain in the Boer War and Malaysia." PhD diss., University of California–San Diego, 1991.

Avant compares how the British successfully adapted to Boer guerrilla military operations in South Africa during the Boer War and Malaysia in the 1950s and 1960s while the U.S. was unable to adapt to Vietnamese communist threats during the Vietnam War. Particular emphasis is placed on the role played by delegated power in civil military relations by these two countries. A key conclusion is that the unified civilian authority in the British Parliament allowed civilian leaders to encourage greater British military doctrinal flexibility through controlling personnel in contrast to the divisive role that civilian policymakers and congressional oversight can play in formulating U.S. military doctrine.

Bickel, Keith B. *Mars Learning: The Marine Corps Development of Small Wars Doctrine, 1915–1940.* PhD diss., Johns Hopkins University, 1999.

Bickel examines Marine Corps counterinsurgency doctrine developed over the Corps's experience fighting small wars in Haiti, the Dominican Republic, and Nicaragua during this time period. A key finding of this work is the role played by low to mid-grade field officers in creating and promoting doctrine, sometimes with opposition from superior officers, which contributed to U.S. success in these conflicts and which led to the 1940 publication of the Marine Corps's *Small Wars Manual.*

Booker, David Lyons. *Cultural Conditioning in Public Organizations: A Survey of the Ideological Perspectives of Air War College Students.* DPA diss., The University of Alabama, 1996.

Using as a benchmark the 1986 Goldwater-Nichols Department of Defense Reorganization Act, which mandated greater inter-service collaboration, this dissertation examines the importance of joint indoctrination on the values of joint specialty officers in the 1993 Air War College class. Mixed findings were gleaned from surveys taken of class members. The majority of respondents saw no basic difference between service doctrine and joint doctrine, while both aviators and non-aviators believed it was very important that basic military doctrine serve as a template for conducting military operations. Class members were evenly divided on whether Air Force doctrine impeded the employment of integrated military power; whether Air Force doctrine could do a better job reflecting the holistic nature of joint military doctrine; and whether they believed joint doctrine should be more general and non-prescriptive, with service doctrine providing specific operational and tactical guidance.

Carlough, Montgomery Cybele. *Pax Brittania: British Counterinsurgency in Northern Ireland, 1969–1982.* PhD diss., Yale University, 1994.

Carlough provides an examination of British counterinsurgency policy against the Irish Republican Army in Northern Ireland. Key findings include emphasizing that while military strategy depends on knowing one's enemy, counterinsurgency warfare promotes beliefs that may involve ethnic vilification of opposing forces. Consequently, military paradigms, even if partially reflective of operational reality, often persist even when they have been demonstrated to fail. This can cause traditional militaries to adopt the tactical and normative practices of their opponents.

Cassidy, Robert Michael. *The Uptonian Paradox and the Cardwellian Conundrum: A Comparison of United States and British Military-Strategic Cultures and Peace Operations Doctrine, 1990–1995*. PhD diss., Fletcher School of Law and Diplomacy at Tufts University, 2000.

Cassidy examines why the U.S. and British militaries, despite possessing so many institutional similarities, have different military doctrines for peace operations. The author sees U.S. peacekeeping operations doctrine as being more forceful in intensity than British doctrine. He also stresses the importance of the Civil War in shaping U.S. military doctrine and the importance of imperial policing responsibilities in sculpting British military doctrine. Contrasts between these two countries during the 19th century involve the U.S. military fighting counterinsurgency operations against American Indians while striving to emphasize Civil War and European military models, with the British military exhibiting greater skill in non-Western military operations and not performing as well in conventional conflicts such as the Crimean War.

A more recent area of emphasis on these countries' peacekeeping doctrine is the role of U.S. experiences in Somalia and British experiences in Bosnia during the early 1990s. A key result of these experiences is U.S. insistence on strong forces, robust rules of engagement, and U.S. command as essential preconditions for U.S. involvement in peacekeeping operations. The author also asserts that the British military's regimental system might be better suited for peacekeeping operations due to its ability to flexibly adapt to evolving, on-the-ground realities.

Corum, James Sterling. *The Reichswehr and the Concept of Mobile War in the Era of Hans von Seeckt*. PhD diss., Queens University at Kingston (Canada), 1990.

This work examines how the German Army examined organizational, tactical, and technical lessons from World War I and used these insights to create an effectual and comprehensive mobile warfare doctrine that would serve as the cornerstone for World War II's blitzkrieg tactics. Development of this doctrine lead the Germans to rebuild and retrain their entire army and develop weapons systems to implement this new doctrinal posture under the leadership of individuals such as Colonel General Hans von Seeckt (1866–1936). Subsequent revision of this dissertation would see its publication as *Hans von Seeckt and German Military Reform* by the University Press of Kansas in 1992.

Cote, Owen Reid, Jr. *The Politics of Innovative Military Doctrine: The United States Navy and Fleet Ballistic Missiles*. PhD diss., Massachusetts Institute of Technology, 1996.

Cote analyzes the roles played by the Polaris and Trident submarine-launched ballistic missile in U.S. Navy strategic nuclear force modernization. Polaris and Trident were developed due to U.S. concerns that its Air Force land-based nuclear systems were vulnerable to Soviet attack. Both of these Navy systems provided a superior alternative to the existing bomber and ICBM systems, but only Polaris produced innovative U.S. nuclear doctrinal changes. Service branch rivalry played some role in developing these two systems and Cote argues that civilian defense leaders can exploit inter-service competition to produce doctrinal innovation.

Edwards, Britt Lynn. *Reforming the Army: The Formulation and Implementation of "Airland Battle 2000."* PhD diss., University of California, Santa Barbara, 1985.

This work examines the military's 1980s' attempt to reform U.S. military doctrine from emphasizing attrition-style warfare to a maneuver-based doctrinal philosophy called AirLand Battle 2000. This work examines the factions in the Pentagon that support and oppose AirLand Battle. "Mavericks" believe it is fraudulent and intended to protect the army's share of the defense budget. "Techno-freaks" are described as enthusiastic supporters. "Moderates" laud its doctrinal reforms but assert that visions of an electronic battlefield may displace decentralized operations. "Hegelians" believe that ongoing military technology advances will frustrate attempts to comprehensively prescribe military policy. Edwards believes the "Moderates" have the strongest critique. This work also examines reactions to AirLand Battle from congressional and European sources.

Farley, Robert M. *Transnational Determinants of Military Doctrine.* PhD diss., University of Washington, 2004.

This work stresses that military doctrine is a critical component of military organization and that these organizations learn doctrine through collaboration. This collaboration results in knowledge sharing, which is critical in developing and executing military doctrine.

Farley examines three case studies of transnational military cooperation, focusing on German-Soviet military cooperation from 1921–1941, U.S. Navy and Royal Navy cooperation from 1914–1945, and U.S. Army and Israeli Defense Force cooperation from 1948–2001.

His research reveals that some mutual absorption of collaborators' doctrinal practices occurred, but that national military doctrinal needs and interests would often be retained by individual participants. It also stresses the importance of civil-military relations in influencing military doctrine, while suggesting possible approaches the United States may want to follow in seeking to develop Afghan and Iraqi militaries capable of defending their countries against internal and external threats.

Foisy, Cory A. *Soviet War-Readiness and the Road to War: 1937–1941.* Master's thesis, McGill University, 2004.

This thesis examines Soviet foreign and domestic policies pertinent to its war-readiness. Key sections include discussion of Soviet industrialization and industrial war preparations between 1928 and 1941; the development of Soviet military doctrine before and after the June 12, 1937, arrest of Marshal Mikhail Tukhachevsky, which initiated Stalin's purge of the Soviet military; examination of how military administrative changes in the late 1930s may have negatively affected initial Soviet performance during the war; and review of Soviet foreign policy during the four years prior to the war. The conclusion asserts that Soviet industrial accomplishments during this period facilitated their successful resistance against the German onslaught.

Gillespie, Paul G. *Precision Guided Munitions: Constructing a Bomb More Potent than the A-Bomb.* PhD diss., Lehigh University, 2002.

Gillespie examines the development of precision-guided munitions whose origins derive from the two world wars. Their development in the late 1960s came about from technical advances in fields such as lasers and semiconductor integrated circuits and as a result of collaboration between the U.S. military, federal government, and civilian industry. Gillespie contends these weapons came about in response to American societal ethics and values, which sought to limit collateral damage and casualties in military operations.

This has produced a U.S. military doctrine that places high emphasis on using precision guided munitions with mixed results. This dissertation would eventually be published as the book, *Weapons of Choice: The Development of Precision Guided Munitions,* by the University of Alabama Press in 2006.

Guttieri, Karen. *Toward a Usable Peace: United States Civil Affairs in Post-Conflict Environments.* PhD diss., University of British Columbia, 1999.

This work is an assessment of U.S. attempts to establish in-country political objectives following military interventions in the Dominican Republic (1965), Grenada (1983), and Panama (1989). These interventions received inconsistent political guidance from Washington and failed to satisfactorily plan for civil political administration. Factors influencing the U.S. approach to civil affairs in these cases included the analytical reasoning behind these interventions, existing orientation toward low-intensity conflict at the time of the intervention, the impact of combat operations during these interventions, and local resource reconstruction availability following the intervention.

The increasing involvement of multiple nations and government agencies in such post-conflict environments increases cultural tensions and makes civil policy efforts more complex.

Hays, Peter Lang. *Struggling Towards Space Doctrine: U.S. Military Space Plans, Programs, and Perspectives during the Cold War.* PhD diss., Fletcher School of Law and Diplomacy at Tufts University, 1994.

This study examines how U.S. military thinking about space and national security evolved during the Cold War era. Hays divides this era into four periods: 1945 to Sputnik in 1957; from Sputnik to 1963; 1964 to 1978; and 1979 to 1989. Questions examined in this treatise include whether national security considerations or organizational behavior input in developing U.S. military space doctrine was more critical during the Cold War; what were the most prominent U.S. military doctrinal tenets during these periods, and how did they relate to U.S. space policy; what were the specific relationships between individual

U.S. military space organizations and specific military space doctrine beliefs; and whether space power's Cold War development path is following the airpower development path that resulted in the 1947 creation of an independent Air Force.

Hays finds that national security considerations tended to be more important than organizational behavioral inputs in conditioning Cold War military space doctrine; that doctrinal issues had a major impact on the creation and preferences of military space organizations; and that airpower's historical development is inappropriate for describing Cold War space power development.

Hayward, Daniel John. *The Operational Manoeuvre Group in Soviet Military Doctrine.* Master's thesis, Carleton University, 1987.

Hayward provides a Canadian examination of the role played by the Red Army's tank and mechanized mobile groups (operational maneuver group) (OMG), which were developed to enable Soviet forces to fight and win conventional wars without escalating to nuclear war. Hayward's work is divided into three parts: analysis of this group's history and operations in Soviet military doctrinal framework; analysis of the vulnerable points of the OMG concept and the Soviet Army's ability to implement it; and evaluation of the effectiveness of NATO strategy to cope with a Warsaw Pact offensive featuring these groups.

The author concludes that OMG could be a potentially useful supplement to Soviet strategy that could significantly assist Soviet offensives in Central Europe, although it could not achieve a deep penetration of NATO defenses or prevent nuclear retaliation; that it would destabilize the Central European balance of power by strengthening Soviet doctrine to rapidly defeat NATO; that it will force NATO to defend against attacks by Warsaw Pact airborne and heliborne troops and special forces; and that OMG destruction of NATO nuclear assets would likely lead to the nuclear escalation OMG seeks to avoid.

Johnson, Wray Ross. *From Counterinsurgency to Stability and Support Operations: The Evolution of United States Military Doctrine for Foreign Internal Conflict, 1961–1996.* PhD diss., Florida State University, 1997.

This dissertation examines how U.S. military doctrine has evolved to encompass concepts such as "counterinsurgency," "low intensity conflict," "military operations other than war," and "stability and support operations" in the time period covered within this work. Subjects examined include the history and nature of irregular warfare, discussion of revolutionary and counterrevolutionary theory, examination of counterinsurgency doctrine's development during the early 1960s, and the strategic context in which these doctrines emerged and how they have been analyzed.

His analysis of foreign conflicts involving U.S counterinsurgency military doctrine includes Vietnam, El Salvador, Somalia, and Bosnia-Herzegovina. In his conclusion, Johnson notes that the United States' recent record in ending internal conflicts has been poor and that U.S. military doctrine supporting conventional military conflicts tends to remain preeminent in the minds of military policymakers. This occurs because American cultural values tend to support the promotion of democratization and hold the belief that there can be solutions to foreign internal conflicts when such solutions don't exist.

An additional limitation on U.S. counterinsurgency military doctrine is that Americans tend to like decisive military victories and are uncomfortable with stalemate and ambiguity when conducting military operations.

Kier, Elizabeth. *Changes in Conventional Military Doctrines: The Cultural Roots of Doctrinal Change*. PhD diss., Cornell University, 1992.

Kier examines the roles played by civilian and military actors as well as unique British and French cultural factors in developing military doctrine. Examples of these unique cultural factors include the high casualties at the World War I Battle of Verdun, which influenced subsequent French military doctrine to emphasize the importance of entrenched defenses, as embodied by the Maginot Line, and how British military doctrine at the outbreak of World War II saw the Royal Air Force focused on confronting a German aerial assault while the British Army nostalgically focused on meeting its multifaceted imperial needs.

Kier maintains that interaction between domestic political arena constraints and military organizational cultural constraints helps determine offensive and defensive military doctrine choices. She goes on to assert that civilians endorse military policy options they believe will maintain existing domestic power levels and that these civilian choices drastically constrain the military's organizational perception of its flexibility in adopting what it regards as desirable doctrinal orientations.

Kilcullen, David J. *The Political Consequences of Military Operations in Indonesia, 1945–1999*. PhD diss., University of New South Wales–Australian Defence Force Academy, 2000.

This work examines the political effects of low-intensity warfare in Indonesia since 1945. Its author is a former Australian military officer and a prominent advisor to the U.S. military and State Department on Iraq. His assessment stresses that analysis of insurgent movements indicates that guerilla group power structures tend to be regionalized and focused on multiple centers of roughly equal authority. He also argues that successful counterinsurgency (COIN) depends on effective political control over the local population, which is generally exercised by local or regional military commanders instead of by centralized authority.

Specific examples of internecine Indonesian warfare examined here include the Darul Islam insurgency in West Java from 1948–1962 and campaigns in East Timor from 1974–1999. Factors influencing these crises included pressures within Indonesian society caused by modernization and other changes from traditional hierarchies to modern social organizational forms; Japan's World War II invasion of what is now Indonesia; the Cold War; Asian financial crises; and increasing economic and media globalization.

Principal findings of the conflicts analyzed here indicate that command and control structures characterizing traditionally dispersed rural guerilla movements lack access to mass media or electronic communications and generally reduce the ability of central political or military leaders to control these movements. Implementing COIN measures, however, will increase the ability of local military leaders to control the civilian population at the expense of other local or central political leaders, and pyramidal or segmented military command structures will result in local commanders having increased authority. Additionally, informal power structures within these societies will be determined by geography, political culture, traditional authority patterns within the society, and how much interaction systemic/regional factors have with local events, all of which will also influence the outcome of COIN operations.

Kinahan, Graham McKnight. *Indian Military Doctrine, 1960–1990.* PhD diss., Fletcher School of Law and Diplomacy at Tufts University, 1994.

This study examines Indian military doctrine from the 1960s–1980s, emphasizing methodologies by reviewing army, navy, and air force orders of battle, senior military personnel assessments of national military capabilities and intentions, and analysis of military training exercises and wartime performance.

Kinahan finds that Indian armed services have been more successful in getting military spending for resources and doctrine that emphasize offensive military operations. While defense spending was maintained at approximately four percent of the Gross Domestic Product, the defense budget and the national economy grew steadily, which enabled India to afford and purchase significant quantities of relatively inexpensive but increasingly sophisticated Soviet weaponry.

These expenditures were intended to deter potentially hostile action by China and Pakistan, boost India's international prestige and its ability to strive for regional strategic dominance, conduct overseas military operations, manipulate foreign threats, and stifle or distract domestic political opponents.

MacDonald, Christian W. *Picking up the Pieces: The Johnson Administration and the Changing Orientation of NATO, 1963–1968.* Master's thesis, University of New Brunswick, 1999.

This appraisal examines policy decisions taken by the Johnson Administration to cope with the North Atlantic Treaty Organization's (NATO) changing strategic environment during the 1960s. A key assertion of this study is that Johnson Administration NATO policy shifted from the hegemonic U.S. policy pursued by the Kennedy Administration to the more multilateral conception of NATO adhered to by the Eisenhower Administration. This policy shift was accelerated by France's decision to withdraw from NATO's military command structure, which made multilateral cooperation critical for future NATO success; the August 1968 Soviet invasion of Czechoslovakia, which ended decreasing western defense budgets and silenced U.S. congressional advocates of withdrawing from NATO; and by ongoing U.S. problems fighting the Vietnam War.

Mallett, Ross A. *Australian Army Logistics 1943–1945*. PhD diss., University of New South Wales–Australian Defence Force Academy, 2007.

This analysis reviews logistical support of the Australian Army's operations in the southwest Pacific from January 1943–August 1945. It opens by examining the overall regional strategic context, with subsequent chapters covering doctrine, base development, storage and tropic proofing issues, inland water transport, air supply, amphibious operations, and combat operations support.

Mallett's concluding assessments stress the critical importance of logistics in supporting Australian as well as American operations in this theatre of war; the necessity of having dependable air, road, and water transport capabilities; and the need for Australia to have the doctrinal flexibility to transform its capabilities from a European-style conflict to the requirements of war fighting in a tropical area that would lead to Australian and allied victories.

Nichols, Thomas Michael. *The Politics of Doctrine: Khrushchev, Gorbachev and the Soviet Military*. PhD diss., Georgetown University, 1988.

This dissertation illuminates the political foundations of Soviet military doctrine, its origins, and what these reveal about Soviet attitudes toward international conflict and Soviet politics. Particular emphasis is placed on military doctrine from 1959–1964 under Khrushchev and from 1986–1988 under Gorbachev.

Nichols maintains that military doctrine is shaped by domestic politics and global events. He contends that external changes such as technological changes or evolving Western attitudes and policies can serve to initiate Soviet doctrinal debate, with tension between civilian and military elites creating conflicting situations that increase rhetorical severity between these elites. Civilian seizure of military doctrine, based on desires to reflect foreign policy goals and enhance civilian control of the military, was an issue that saw the General Secretary and his cohorts contend that they should have an almost exclusive prerogative to formulate military doctrine, which was resisted by the Soviet military on national security and political grounds.

Petraeus, David Howell. *The American Military and the Lessons of Vietnam: A Study of Military Influence and the Use of Force in the Post–Vietnam Era*. PhD diss., Princeton University, 1987.

This dissertation, written by the current commander of U.S. and coalition military forces in U.S. Central Command, examines the impact of Vietnam on America's senior military in terms of advising national political leadership on using American military forces in potential combat situations. It argues that the post–Vietnam U.S. military has been extremely cautious in advocating the use of force, and there is no example of military leadership offering more aggressive recommendations to use force than the most hawkish civilian advisors.

Petraeus adds that caution is likely to characterize military attitudes toward using force for some time, but that this caution from Vietnam experiences may be ambiguous and overlook potential problems such caution may present when dealing with emerging national security threats.

Prunckun, Henry Walter, Jr. *Operation El Dorado Canyon: A Military Solution to the Law Enforcement Problem of Terrorism.* Master's thesis, University of South Australia, 1995.

Prunckun studies the effectiveness of Operation El Dorado Canyon, which was the U.S. air raid on Tripoli, Libya in April 1986. This action was launched in retaliation for Libyan support for international terrorism. He maintains that the Reagan Administration's decision to bomb Libya stemmed from the military doctrine of deterrence, which is a strategy to contain state aggression through the fear of retaliation. Prunckun goes on to assert that counterterrorism had previously been conducted by law enforcement and intelligence agencies and that this was the first time the U.S. military had been used to resolve what had been seen as a law enforcement problem.

As part of its methodology, this work presents three possible outcomes to this attack: a reduction of Libyan-sponsored terrorist attacks; a reduction of terrorist attacks against U.S. citizens and property; or no increase in the number or severity of international terrorist attacks. The principal conclusion from this analysis is that there was a strong relationship between this operation and a decline in terrorist attacks against U.S. targets afterward.

Richardson, Wade (Trey) Franklin, III. *The Gulf War Syndrome Debate: Science, Politics, and the Reshaping of Military Doctrine.* Master's thesis, University of Louisville, 2006.

Gulf War Syndrome refers to the situation in which veterans returning from the 1990–1991 Persian Gulf War suffered from illnesses doctors had difficulty identifying, although there was some belief that these illnesses occurred from exposure to hazardous materials. Richardson shows how awareness of this syndrome initially occurred in the media, prompting veteran activism, medical research, political activism, and a study of Gulf War military records and doctrine. His work reveals how veteran activism stimulated study that succeeded in revising flawed doctrine. At the same time, this veteran activism ultimately had a negative effect on medical research into these illnesses because of veterans' biases against such research that included stress as a possible source of veteran illnesses.

Salazar, Edward Joseph. *Soviet Strategic Doctrine: The Development of a Strategic Concept for External Force Projection.* PhD diss., Claremont Graduate University, 1983.

This work examines the power projection capability or "external force function" of the Soviet military in Soviet doctrinal thought. Salazar contends that there is a hierarchy of thought level within Soviet military thinking on doctrine, strategy, science, and art. Soviet doctrine represents a complex, political-economic amalgamation that determines overall policy according to future warfare needs. He also maintains that Marxism-Leninism provides military leaders with an evaluative framework for assessing military policy requirements.

An increasing need for the ability to project power globally contributed to a doctrinal shift from emphasizing global nuclear war to conventional local war, which was reflected by the phrase, "national liberation," becoming an important part of Soviet strategic doctrine. Consequently, Soviet military doctrine and strategic objectives depend on appropriate use of military power.

Searle, Dean. *Low Intensity Conflict: Contemporary Approaches and Strategic Thinking.* PhD diss., University of Waikato, 2007.

This New Zealand work examines how low intensity conflict (LIC) has become a significant feature of contemporary military conflict and how such conflict poses particular challenges for conventional armed forces that are likely to increase in the future. Searle's primary focus is on practical aspects of ending LIC, with a key research emphasis on establishing a doctrinal and military framework to prevent and resolve LIC.

He examines Russian experiences with LIC conflict in Afghanistan and Chechnya, U.S. experiences in Somalia and Afghanistan, American and British experiences in Iraq during 2003–2004, general principles for using military force in LIC, and Australian and New Zealand experiences with LIC.

His conclusions stress the necessity of taking a holistic approach to LIC counterinsurgency operations; the need for specific and comprehensive doctrine to suppress insurgencies; developing customized strategies to counter organizational, terrorist, guerilla, and mobile warfare phases of insurgencies; and being able to control international interference in such conflicts in order restore civil order, which also requires winning the support of the civilian population.

Twomey, Christopher P. *The Military Lens: Doctrinal Differences, Misperception, and Deterrence Failure in Sino-American Relations.* PhD diss., Massachusetts Institute of Technology, 2005.

Twomey believes that nations have divergent strategic situations, histories, and military cultures, which combine to produce variant beliefs about effective military doctrine, strategy, and capabilities. He argues that when such doctrines of military victory theories differ between states, misperceptions and false optimism are likely to occur. Such misperceptions may restrict international diplomacy by making communication and common balance of power assessments more difficult, which may result in conflict escalation and war.

This treatise examines scholarship on military doctrine sources, strategic culture, misperception, strategic coercion, and deterrence theory. Particular emphasis is placed on strategic coercion attempts in early Cold War Sino-American conflicts in Korea and the Taiwan Strait. Emphasis is placed on how communication between each of these powers depended on their own doctrinal theories of victory, which, Twomey believes, impeded diplomatic activity between China and the United States. A key conclusion is that policymakers need to carefully review perceptual frameworks of military doctrine held by policymakers they are trying to influence.

Van Nort, Richard M. *The Battle of Adrianople and the Military Doctrine of Vegetius.* PhD diss., City University of New York, 2007.

Dissertation examining the relationship between a Roman military defeat against the Goths at Adrianople around 376 AD and a document within the following century called "De Rei Militari," written by Flavius Vegetius Renatus and presented to the Roman Emperor. "De Rei Militari" states that Rome was mistaken in allowing its heavily armored infantry to deteriorate and that it was possible to correct this situation by returning to traditional Roman military practices.

According to Van Nort, "De Rei Militari" called for close cavalry and light infantry collaboration, the necessity for light cavalry to perform reconnaissance and screening functions, the need for protracted combat against enemies with long supply lines and at great distances from their homes, and protecting Roman cities, towns, and roads by fortifying them. Vegetius' work would ultimately be used by Roman and later Byzantine military forces in subsequent centuries to emphasize the importance of both cavalry and infantry in meeting emerging infantry and cavalry threats from forces such as the Goths and Mongols.

Waddell, Timothy Scott. *Marshal N.V. Ogarkov and the Transformation in Soviet Military Affairs.* Master's thesis, University of Manitoba, 1999.

Waddell examines how Marshal Nikolai Ogarkov (1917–1994) interpreted and contributed to Soviet military doctrine while serving as Soviet armed forces Chief of General Staff and First Chief Deputy Minister of Defense from 1977 to 1984. Waddell begins by examining Soviet military doctrine from 1917 to 1977 and investigating how the emergence of nuclear weapons shaped Soviet doctrinal thought.

Ogarkov possessed a strong understanding of Marxist theory, which influenced his military doctrinal views. These included the importance of technological change in shaping conventional and military doctrine; the replacement of cavalry by mechanized forces; his belief in the primary importance of military doctrine's socio-political aspects; that strategy must be subordinated to military doctrine; his belief in the early 1980s that the U.S. was striving for military superiority over the Soviet Union; and his concern that the technological revolution would be militarily injurious to the Soviet Union.

Waddell believes Ogarkov was more successful than his predecessors in turning his ideas on military strategic and technological change into military doctrinal and operational reality. Key examples of Ogarkov's ideas in this regard were his apparent rejection of military doctrine relying on nuclear weapons for victory, accepting western belief in mutually assured destruction, rejecting the limited use of nuclear weapons in war, and believing that destructive nature of nuclear weapons negates warfare. This, in turn, caused Ogarkov to place increasing emphasis on the growing importance of high-technology conventional weapons in Soviet military doctrinal thought.

Zisk, Kimberly Marten. *Soviet Reactions to Shifts in U.S. and NATO Military Doctrine in Europe: The Defense Policy Community and Innovation.* PhD diss., Stanford University, 1991.

Zisk presents an argument against the theory that military institutions resist doctrinal innovation and that civilian intervention is required to overcome such resistance. Instead, military officers are inherently reactive to foreign military threats, including hostile doctrinal changes, and such officers prefer adopting reactive doctrinal innovations to counter such threats.

She regards defense experts as individual policy community members instead of the representatives of institutional interests, and concludes that as the defense policy community changes or expands, either through military or personnel turnover or through the influx of newly empowered civilian experts, it will be easier to incorporate doctrinal innovations since new community members are less likely to adhere to the status quo.

Three case studies are presented in this dissertation: Soviet reactions to western adoption of Flexible Response doctrine during the 1960s; American adoption of the Schlesinger doctrine in 1974; and the combined U.S. adoption of AirLand Battle doctrine in 1982 and NATOs 1984–1985 adoption of Follow-On Forces Attack doctrine.

Technical Reports

Technical reports from government agencies such as the National Technical Information Service (NTIS) and the Defense Technical Information Center's (DTIC) Scientific and Technical Information Network (STINET) can also be useful resources for those conducting research on grey literature concerning military doctrine and other scientific and technological subjects. NTIS is a U.S. Commerce Department agency whose purpose is providing and simplifying access to the multitudinous data files and scientific and technical reports produced by federal agencies and their contractors. Initially established as the Publications Board in 1945 to manage the release of captured German documents and technical reports to U.S. industry, NTIS received its present name in 1970. Since NTIS receives no congressional appropriations, it charges for costs associated with collecting, abstracting, storing, reproducing, and selling its information resources through public sales.[6]

NTIS's Web site (http://www.ntis.gov/) provides information about its products and services and how to search for and locate these items. NTIS's Homeland Security Information Center (http://www.ntis.gov/hs/) features a searchable collection of military manuals and information on accessing more than three million titles NTIS possesses. NTIS materials are useful if you desire to purchase copies of military doctrine publications. Many of these resources are available freely elsewhere as has been described in this book, with DTIC STINET being a prime example.

DTIC began after World War II due to the need to translate captured German and Japanese military, scientific, and technical information. The Secretaries of the Navy and Air Force formally established it as the Central Documents Office on October 13, 1948, and it became known as DTIC in October 1979. DTIC serves as a specialized provider of domestic and international scientific and technical reports, with particular emphasis on those having military applications for the Defense Department. General information about DTIC and its products and services is available at http://www.dtic.mil.[7]

STINET (http://stinet.dtic.mil/) provides the ability to search for and retrieve abstracts and the full text of many technical reports on military doctrine and other topics. Reports that are not available in full text may be ordered through NTIS. Examples of these reports, many produced by students at military war colleges or research institutes like the Rand Corporation and Institute for Defense Analyses, include the following citations with Uniform Resource Locators.

Brian Manthe, *United States Military Doctrine and the Conduct of Counterinsurgency Operations: Fixing the Disconnect,* http://handle.dtic.mil/100.2/ADA393508 (2001); William A. Forkner, Thomas L. Kelly, and Richard S. Lamarre, *Transformation Déjà Vu?: A Comparison of Military Improvements of Israel (1967–1973) and the United States (1990–2002),* http://handle.dtic.mil/100.2/ADA421638 (2002); Thomas Michael LaMeur, *Mikhail Frunze and the Unified Military Doctrine,* http://handle.dtic.mil/100.2/ADA429032 (2004); Stephen D. Pomper, *Asymmetric: Myth in United States Military Doctrine,* http://handle.dtic.mil/100.2/ADA428994 (2004); David A. Kummings, *Rising China and the ASW Problem,* http://handle.dtic.mil/100.2/ADA470759 (2007); Scott Neitzel, *The Falklands War: Understanding the Power of Context in Shaping Argentine Strategic Decisions* http://handle.dtic.mil/100.2/ADA474391 (2007); Jason D. Ross, *Forcing Doctrine to Match Reality: Bridging the Foreign Military Training Doctrine Gap Within the Australian Defence Force* http://handle.dtic.mil/100.2/ADA475650 (2007); and Charles J. Dunlap Jr., *Shortchanging the Joint Fight: An Airman's Assessment of FM 3–24 and the Case for Developing Truly Joint COIN Doctrine,* http://handle.dtic.mil/100.2/ADA475650 (2008).

Think Tanks

Research institutions or think tanks can also be producers of military doctrine research and analysis. Experts from these organizations may be hired by government departments and military services to conduct research or design projects, and many of them may be invited to testify before congressional committees in support of or opposition to particular legislative proposals. Funding for these institutions may come from individual, nonprofit, governmental, and commercial sources and think tanks, which represent a variety of ideological or philosophical perspectives.[8]

A particularly important think tank for national security policy research and military doctrine research and analysis is the Rand Corporation. It began operations in December 1945 as Project RAND with the initial involvement of the Army Air Force and Douglas Aircraft Company. On May 14, 1948, RAND was incorporated as a nonprofit California corporation with an institutional mission emphasizing the promotion of scientific, educational, and charitable purposes, along with U.S. public welfare and national security.[9]

Rand's Web site (http://www.rand.org/) features a tremendous variety of reports on national security topics, including military doctrine, with many of these reports being in full text. Many academic research libraries receive Rand publications on standing order or have significant numbers of these publications in their collections. Examples of historic Rand military doctrine analyses that are not available on the Internet but that may be available in library collections or purchased include Alice Langley Hsieh, *Communist China's Military Policies, Doctrine, and Strategy: A Lecture Presented at the National Defense College, Tokyo, September 17, 1968;* S. T. Cohen, *On the Stringency Criteria for Battlefield Nuclear Operations* (1975); Robert L. Perry, *The Interaction of Technology and Doctrine in the USAF* (1979); Michael Checinksi, *A Comparison of the Polish and Soviet Armaments Decisionmaking Systems* (1981); Benjamin Lambeth, *Conventional Forces for NATO* (1987); Michael E. Thompson, *Political and Military Components of Air Force Doctrine in the Federal Republic of Germany and Their Implications for NATO Defense Policy Analysis* (1987); Sally W. Stoecker, *Historical Roots of Contemporary Debates on Soviet Military Doctrine and Defense* (1992); Jennifer Taw and Robert C. Leicht, *The New World Order and Army Doctrine: The Doctrinal Renaissance of Operations Short of War?* (1992); and C. Christine Fair, *Military Operations in Urban Areas: The Indian Experience* (2003).

A much greater number of recent Rand reports on military doctrine are accessible on the Rand Web site. Representative samples of these reports with URLs include Patrick D. Allen, *The Pace of War in Gaming, Simulation, Doctrine, and War,* http://www.rand.org/pubs/papers/P7229/ (1986); Mark A. Lorell, *Airpower in Peripheral Conflict: The French Experience on Africa,* http://www.rand.org/pubs/reports/R3660/ (1989); Elwyn Harris, Kenneth Horn, Edison Cesar, and Paul Steinberg, *Recommended Strategy for the Army's Role in Space,* http://www.rand.org/pubs/notes/N3535/ (1993); Eugene Rumer, *The Building Blocks of Russia's Future Military Doctrine,* http://www.rand.org/pubs/monograph_reports/MR359/ (1994); Russell Glenn, *Marching Under Darkening Skies: The American Military and the Impending Urban Operations Threat,* http://www.rand.org/pubs/monograph_reports/MR1007/ (1998); John Arquilla and David Ronfeldt, *Swarming and the Future of Conflict,* http://www.rand.org/pubs/documented_briefings/DB311/ (2000); Dana J. Johnson and Ariel E. Levite, eds., *Toward Fusion of Air and Space: Surveying Developments and Assessing Choices for Small and Middle Powers,* http://www.rand.org/pubs/conf_proceedings/CF177/ (2003); Robert C. Owen and Karl P. Mueller, *Airlift Capabilities for Future U.S. Counterinsurgency Operations,* http://www.rand.org/pubs/monographs/MG565/ (2007); Thomas S. Sazayna, Derek Eaton, and Amy Richardson, *Preparing the Army for Stability Operations: Doctrinal and Interagency Issues,* http://www.rand.org/pubs/monographs/2007/Rand_MG646.pdf (2007); and Paul K. Davis, Russell D. Shaver, and Justin Beck, *Portfolio-Analysis Methods for Assessing Capability Options,* http://www.rand.org/pubs/monographs/MG662/ (2008).

A sampling of other think tanks producing freely available military doctrine research and analysis includes the American Enterprise Institute (http://www.aei.org/); Center for Strategic and Budgetary Assessments (http://www.csbaonline.

org/); Center for Strategic and International Studies (http://www.csis.org/); Heritage Foundation (http://www.heritage.org/); Institute for Defense Analysis (http://www.ida.org/); and many other national security policy-oriented think tanks in the United States and elsewhere. Web sites serving as good directories of think tanks include the University of Michigan Library's Political Science Resources (http://www.lib.umich.edu/govdocs/psthink.html); the Foreign Policy Research Institute Think Tank Directory (http://thinktanks.fpri.org/), and Purdue University Libraries Research Center (http://www.lib.purdue.edu/hsse/infopages/subjectlinks/researchcenters.html).

Conference Proceedings

Conference proceedings can also be useful sources for finding information on military doctrine. Professional associations representing a variety of disciplines hold conferences on a regular basis where members discuss and debate trends and developments in their fields and present their findings and data in speeches, presentations, and, in some cases, through published papers. Some conference proceeding documents may eventually be published as scholarly journal articles or chapters in books. There are numerous assessments in library and information science literature on the role of conference proceedings in scholarly research and communication, and the challenges in accessing these materials.[10]

Most conference proceedings are not freely available to users who are not part of the professional associations in question or affiliated with a university with a major academic library. These resources tend to be selectively or sporadically cataloged in academic library online public access catalogs and often are not cataloged with as high a level of bibliographic access as books and journals.[11]

Two major commercial databases for accessing conference proceedings that are available in some academic libraries include the Institute for Scientific Information's *ISI Web of Knowledge: Proceedings* and Cambridge Scientific Abstracts' *Conference Papers Index*. General information about these resources may be found at http://pcs.isiknowledge.com/ and http://www.csa.com/.

Conference proceedings represent an interdisciplinary variety of subjects. Those covering military doctrine may be produced as part of the scholarly research process in disciplines such as history, political science, military science, and various scientific and technology fields. The following is a selective annotation of relatively recently published conference proceedings on military doctrine topics. Bibliographic citations are provided, including information on the organization at which this paper was initially presented and book International Standard Bibliographic Numbers (ISBN) if available. A representative sampling of these papers arranged in chronological order by the conference date includes:

Levite, A. "Advanced Weaponry, Military Doctrine, and Threat Perceptions in the Middle East." Paper presented at the AAAS 93–159th National Meeting of the American Association for the Advancement of Science, Boston, MA, February 11–16, 1993.

Houchin, R.F. "Doctrine and Dyna-Soar: Origins of USAF Manned Military Spacecraft." Paper presented at the 31st History Symposium of the International Academy of Astronautics, Turin, Italy, October 6–10, 1997. ISBN: 087703-518-0.

Metallinos, P. "The Military and Geostrategic Dimensions of the Truman Doctrine." In *Greece's Pivotal Role in World War II and its Importance to the U.S. Today,* ed. Eugene T. Rossides (2001), 156–165. Paper presented at the American Hellenic Institute Foundation Conference on Greece's Pivotal Role in World War II and its Importance to the U.S. Today, Washington, DC, November 22, 1997. ISBN: 1-889247-03-0.

Huang, A.C.C. "Transformation and Refinement of Chinese Military Doctrine: Refection and Critique on the PLA's View." In *Seeking Truth from Facts—A Retrospective on Chinese Military Studies in the Post–Mao Era,* eds. James C. Mulvenon and Andrew Wang (2001), 131–140. Paper presented at the Meeting on a Retrospective on Chinese Military Studies in the Post–Mao Era, Washington, DC, July 8–11, 1999. ISBN: 0-8330-2936-3.

Yue, Y., B. Kirby, and R. S. Seymour. "Developing an Operational Architecture for the Australian Army Enhanced Combat Force in the Digitised Network-Centric Battlespace." In *Battlespace Digitization and Network-Centric Warfare,* ed. Raja Suresh (2001), 87–98. Paper presented at the 6th Battlespace Digitization and Network-Centric Warfare Conference, Orlando, FL, April 18–20, 2001. ISBN: 0-8194-4091-4.

Cosido. I. "Creating Asymmetric Doctrine: The Role for Security Forces of a Military Nature." In *Future NATO Security: Addressing the Challenges of Evolving Security and Information Sharing Systems and Architectures,* eds. Martin Edmonds and Oldrich Cerny (2004), 119–120. Paper presented at the NATO Advanced Research Workshop on Future NATO Security, Prague, March 8–10, 2003. ISBN 1-586093-392-1.

Bolia, R. S., W. T. Nelson, M. A. Vidulich, and R. M. Taylor. "From Chess to Chancellorsville: Measuring Decision Quality in Military Commanders." In *Human Performance Situation Awareness and Automation: Current Research and Trends, Vol. 1,* eds. Dennis A. Vicenzi, Mustapha Moula, and Peter A. Hancock (2004), 269–273. Paper presented at the 2nd Conference on Human Performance, Situation Awareness and Automation, Daytona Beach, FL, March 22–25, 2004. ISBN: 0-8058-5341-3.

Palmarini, M., and J. Rapanotti. "Integrated Development of Light Armoured Vehicles Based on War-Gaming Simulators." In *Enabling Technologies for Simulation Science VIII,* eds. Dawn A. Trevisani and Alex F. Sisti (2004), 244–251. Paper presented at the Conference on Enabling Technologies for Simulation Science VIII, Orlando, FL, April 13–15, 2004. ISBN: 0-8194-5346-3.

Delic, Bozidar. "The Military Aspects of NATO's Aggression against the FRY." In *Kosovo and Methija: Past, Present, Future,* ed. Kosta Mihailovic (2006), 331–? Paper presented at the International Scholarly Meeting on Kosovo and Metohija—Past, Present, and Future, Belgrade, March 16–18, 2006. ISBN: 978-86-70250429-9.

Fisher, M. and M. Syvret. "A NATO Collective Strategy Proposal and Practical Planning and Analysis Experiences from Operations in Afghanistan." In *Cornwallis Group X: Analysis for New and Emerging Societal Conflicts,* eds. Alexander Woodcock and George A.

Rose (2006), 305–321. Paper presented at the 10th Annual Meeting of the Cornwallis Group, Kingston, Canada, March 21–24, 2005. ISBN 1-896551-61-0.

Hidek, Matt. "Military Doctrine and Integrated Intelligence in the City." Paper presented at the 2007 Meeting of the Association of American Geographers, San Francisco, CA, April 17–21, 2007.

Notes

1. For a representative sampling of writing on grey literature and its role in library collections, see Paola De Castro and Sandra Salinetti, "Quality of Grey Literature in the Open Access Era: Privilege and Responsibility," *Publishing Research Quarterly* 20, no. 1 (2004): 4–12; Heather Lehman and Janet Webster, "Describing Grey Literature Again: A Survey of Collection Policies," *Publishing Research Quarterly* 12, no. 1 (2005): 64–72; Rose M. Jackson, "Grey Literature and Urban Planning: History and Accessibility," *Publishing Research Quarterly* 12, no. 1 (2005): 94–104; and Cherifa Boukacem-Zeghmouri and Joachim Schopfel, "Document Supply and Open Access," *Interlending and Document Supply* 34, no. 3 (2006): 96–104.

2. For representative samples, see Calvin James Boyer, *The Doctoral Dissertation as an Information Source* (Metuchen, NJ: Scarecrow Press, 1973); Edward S. Balian, *How to Design, Analyze, and Write Doctoral Research: The Practical Guidebook* (Lanham, MD: University Press of America, 1982); and Peggy L. Maki and Nancy A. Borkowski, eds., *The Assessment of Doctoral Education: Emerging Criteria and New Models for Improving Outcomes* (Sterling, VA: Stylus, 2006).

3. The roles and problems of providing access to theses and dissertations in library collections and the intellectual and bibliographic content of these materials is analyzed by numerous sources, including Jean-Pierre Herubel and Ann Buchanan, "Comparing Materials Used in Philosophy and Political Science Dissertations: A Technical Note," *Behavioral and Social Sciences Librarian* 12, no. 2 (1993): 63–70; Jean-Pierre Herubel and Ann Buchanan, "Profiling Ph.D. Dissertation Bibliographies: Serials and Collection Development in Political Science," *Behavioral and Social Sciences Librarian* 13, no. 1 (1994): 1–10; L. Hoover, "Cataloging Theses and Dissertations: An Annotated Bibliography," *Technical Services Quarterly* 19, no. 3 (2001): 21–39; L. Hoover and R. E. Wolverton, "Cataloging and Treatment of Theses, Dissertations, and ETDs," *Technical Services Quarterly* 20, no. 4 (2003): 3–57; Yale Fineman, "Electronic Theses and Dissertations," *Portal: Libraries and the Academy* 3, no. 2 (2003): 219–227; Newkirk Barnes, "The Use of U.S. Government Publications as Bibliographic References in Doctoral Dissertations," *Journal of Academic Librarianship* 32, no. 5 (2006): 503–511; "Purdue Libraries Launches e-Scholar," *Advanced Technology Libraries* 35, no. 11 (2006): 1, 11–12; William Clark, *Academic Charisma and the Origins of the Research University* (Chicago: University of Chicago Press, 2006), 297–335; Thomas E. Nisonger, "A Review and Analysis of Library Availability Studies," *Library Resources and Technical Services* 51, no. 1 (2007): 30–49; and Eun G. Park, Qing Zou, and David McKnight, "Electronic Thesis Initiative: Pilot Project of McGill University," *Program: Electronic Library and Information Systems* 41, no. 1 (2007): 81–91.

4. For compilations and analysis of military theses and dissertations, see Allan Reed Millett and B. Franklin Cooling III, *Doctoral Dissertations in Military Affairs: A Bibliography* (Manhattan: Kansas State University Library, 1972); L. L. Sims and A. D. Officer, eds., *Abstracts of Theses/Special Studies, 1964–1976: Master of Military Art and Science for 1964–1976*

(Fort Leavenworth, KS: U.S. Army Command and General Staff College, 1976?); Jean-Pierre Herubel and Edward A. Goedeken, "Dissertations in Military History, 1973–1988: A Survey and Analysis," *Journal of Military History* 56 (1992): 651–657; and Edward A. Goedeken and Dennis E. Showalter, "Doctoral Dissertations in Military History," *Journal of Military History* 71 (2007): 1007–1023.

5. See Air University, *AU-10 Air University Catalog: Academic Year 2007–2008* (Maxwell Air Force Base, AL: Air University Press, 2007), 40; U.S. Army War College, *Curriculum Catalogue: Academic Year 2006–2007* (Carlisle Barracks, PA: U.S. Army War College, 2006), 15; U.S. Army Command and General Staff College, *Circular 12–1 Chapter 7* (Fort Leavenworth, KS: U.S. Army Command and General Staff College, 2007), 4–5; National Defense University, *NDU Catalog* (Washington, DC: National Defense University, 2007), 22–29; Naval Postgraduate School, *Academic Catalog* (Monterey, CA: Naval Postgraduate School, 2008), 15; and U.S. Naval War College, College of Naval Command and Staff, *Writing Guide* (Newport, RI: U.S. Naval War College, College of Naval Command and Staff, 2007?).

6. Bert Chapman, *Researching National Security and Intelligence Policy* (Washington, DC: CQ Press, 2004), 75.

7. Lane E. Wallace, *The Story of the Defense Technical Information Center: 1945–1995* (Fort Belvoir, VA: Defense Technical Information Center, 1995), 10–15, 46.

8. For a listing of prominent U.S. national security policy-oriented think tanks and descriptions of their research, see Chapman, *Researching National Security,* 296–326.

9. See Rand Corporation, "About Rand: History and Mission" (2008), 1–7, http://www.rand.org/history (accessed February 29, 2008) and Martin J. Collins, *Cold War Laboratory: RAND, the Air Force, and the American State, 1945–1950* (Washington, DC: Smithsonian Institution Press, 2002).

10. See Kimberly Douglas, "Conference Proceedings at Publishing Crossroads," *Science and Technology Libraries* 22, no. 3/4 (2002): 39–50; James Hartley, "On Requesting Conference Papers Electronically," *Journal of Information Science* 30, no. 5 (2004): 475–479; Yannis Manolopoulos and Antonis Sidiropoulos, "A New Perspective to Automatically Rank Scientific Conferences Using Digital Libraries," *Information Processing and Management* 41, no. 2 (2005): 389–312; and Wolfgang Glanzel, Balazs Schlemmer, Andras Schubert, and Bart Thus, "Proceedings Literature as Additional Data Source for Bibliometric Analysis," *Scientometrics* 68, no. 3 (2006): 457–473.

11. Examples of literature that examine this include Barbara L. Berman, "Coping with Conference Proceedings," *Cataloging and Classification Quarterly* 10, no. 3 (1990): 19–34; J. H. Bowman, "Changing Cataloging Rules in Relation to Changing Patterns of Publication," *Cataloging and Classification Quarterly* 22, no. 2 (1996): 29–50; S. M. DeSilva and A. N. Zainab, *Cataloging and Classification Quarterly* 29, no. 3 (2000): 63–80; and B. M. Russell and R.L.B. Hutchison, "Official Publications at Texas A&M University: A Case Study in Cataloging Archival Material," *American Archivist* 63, no. 1 (2000): 175–184.

Index

About the Author

BERT CHAPMAN is Government Information/Political Science Librarian and Professor of Library Science at Purdue University Libraries. He received his B.A. in history/political science from Taylor University, an M.A. in history from the University of Toledo, and an M.S.L.S. in library science from the University of Kentucky. Prior to his service at Purdue, he was Reference/Documents Librarian at Lamar University. His research interests include using government documents to conduct military policy research and other forms of historical research. He is the author of *Space Warfare and Defense: A Historical Encyclopedia and Research Guide* and *Researching National Security and Intelligence Policy*.